Innovative Applications of Mo(W)-Based Catalysts in the Petroleum and Chemical Industry:

Emerging Research and Opportunities

Hui Ge
Chinese Academy of Sciences, China

Xingchen Liu
Chinese Academy of Sciences, China

Shanmin Wang
Oak Ridge National Laboratory, USA

Tao Yang
China University of Petroleum, China

Xiaodong Wen
Synfuels China, China

A volume in the Advances in Chemical and Materials Engineering (ACME) Book Series

www.igi-global.com

Published in the United States of America by
IGI Global
Engineering Science Reference (an imprint of IGI Global)
701 E. Chocolate Avenue
Hershey PA 17033
Tel: 717-533-8845
Fax: 717-533-8661
E-mail: cust@igi-global.com
Web site: http://www.igi-global.com

Copyright © 2017 by IGI Global. All rights reserved. No part of this publication may be reproduced, stored or distributed in any form or by any means, electronic or mechanical, including photocopying, without written permission from the publisher.
Product or company names used in this set are for identification purposes only. Inclusion of the names of the products or companies does not indicate a claim of ownership by IGI Global of the trademark or registered trademark.
   Library of Congress Cataloging-in-Publication Data

Library of Congress Cataloging-in-Publication Data

Names: Ge, Hui, 1964- author.
Title: Innovative applications of Mo(W)-based catalysts in the petroleum and
   chemical industry : emerging research and opportunities / by Hui Ge,
   Xingchen Liu, Shanmin Wang, Tao Yang, and Xiaodong Wen.
Description: Hershey, PA : Engineering Science Reference, [2017] | Includes
   bibliographical references.
Identifiers: LCCN 2016057549| ISBN 9781522522744 (hardcover) | ISBN
   9781522522751 (ebook)
Subjects: LCSH: Catalysis. | Catalytic reforming. | Molybdenum
   compounds--Industrial applications. | Chalcogenides--Industrial
   applications. | Green chemistry.
Classification: LCC TP156.C35 G423 2017 | DDC 660/.2995--dc23 LC record available at https://lccn.loc.gov/2016057549

This book is published in the IGI Global book series Advances in Chemical and Materials Engineering (ACME) (ISSN: 2327-5448; eISSN: 2327-5456)

British Cataloguing in Publication Data
A Cataloguing in Publication record for this book is available from the British Library.

All work contributed to this book is new, previously-unpublished material. The views expressed in this book are those of the authors, but not necessarily of the publisher.

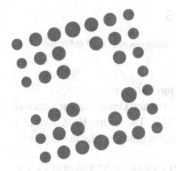

Advances in Chemical and Materials Engineering (ACME) Book Series

ISSN:2327-5448
EISSN:2327-5456

Editor-in-Chief: J. Paulo Davim, University of Aveiro, Portugal

MISSION

The cross disciplinary approach of chemical and materials engineering is rapidly growing as it applies to the study of educational, scientific and industrial research activities by solving complex chemical problems using computational techniques and statistical methods.

The **Advances in Chemical and Materials Engineering (ACME) Book Series** provides research on the recent advances throughout computational and statistical methods of analysis and modeling. This series brings together collaboration between chemists, engineers, statisticians, and computer scientists and offers a wealth of knowledge and useful tools to academics, practitioners, and professionals through high quality publications.

COVERAGE

- Metallic Alloys
- Ductility and Crack-Resistance
- Sustainable Materials
- Composites
- Coatings and surface treatments
- Wear of Materials
- Fracture Mechanics
- Biomaterials
- Artificial Intelligence Methods
- Materials to Renewable Energies

IGI Global is currently accepting manuscripts for publication within this series. To submit a proposal for a volume in this series, please contact our Acquisition Editors at Acquisitions@igi-global.com or visit: http://www.igi-global.com/publish/.

The Advances in Chemical and Materials Engineering (ACME) Book Series (ISSN 2327-5448) is published by IGI Global, 701 E. Chocolate Avenue, Hershey, PA 17033-1240, USA, www.igi-global.com. This series is composed of titles available for purchase individually; each title is edited to be contextually exclusive from any other title within the series. For pricing and ordering information please visit http://www.igi-global.com/book-series/advances-chemical-materials-engineering/73687. Postmaster: Send all address changes to above address. Copyright © 2017 IGI Global. All rights, including translation in other languages reserved by the publisher. No part of this series may be reproduced or used in any form or by any means – graphics, electronic, or mechanical, including photocopying, recording, taping, or information and retrieval systems – without written permission from the publisher, except for non commercial, educational use, including classroom teaching purposes. The views expressed in this series are those of the authors, but not necessarily of IGI Global.

Titles in this Series

For a list of additional titles in this series, please visit:
http://www.igi-global.com/book-series/advances-chemical-materials-engineering/73687

Sustainable Nanosystems Development, Properties, and Applications
Mihai V. Putz (West University of Timişoara, Romania & Research and Development National Institute for Electrochemistry and Condensed Matter (INCEMC) Timişoara, Romania) and Marius Constantin Mirica (Research and Development National Institute for Electrochemistry and Condensed Matter (INCEMC) Timişoara, Rmania)
Engineering Science Reference • ©2017 • 794pp • H/C (ISBN: 9781522504924) • US $245.00

Computational Approaches to Materials Design Theoretical and PracticalAspects
Shubhabrata Datta (Calcutta Institute of Engineering and Management, India) and J. Paulo Davim (University of Aveiro, Portugal)
Engineering Science Reference • ©2016 • 475pp • H/C (ISBN: 9781522502906) • US $215.00

Green Approaches to Biocomposite Materials Science and Engineering
Deepak Verma (Graphic Era Hill University, Dehradun, India) Siddharth Jain (College of Engineering Roorkee, India and University of Alberta, Canada) Xiaolei Zhang (Queens University, Belfast, UK) and Prakash Chandra Gope (College of Technology, G.B.Pant University of Agriculure and Technology, Pantnagar, India)
Engineering Science Reference • ©2016 • 322pp • H/C (ISBN: 9781522504245) • US $165.00

Position-Sensitive Gaseous Photomultipliers Research and Applications
Tom Francke (Myon, Sweden) and Vladimir Peskov (CERN, Switzerland)
Engineering Science Reference • ©2016 • 562pp • H/C (ISBN: 9781522502425) • US $240.00

Research Perspectives on Functional Micro- and Nanoscale Coatings
Ana Zuzuarregui (CIC nanoGUNE Consolider, Spain) and Maria Carmen Morant-Miñana (CIC nanoGUNE Consolider, Spain)
Information Science Reference • ©2016 • 511pp • H/C (ISBN: 9781522500667) • US $215.00

For an enitre list of titles in this series, please visit:
http://www.igi-global.com/book-series/advances-chemical-materials-engineering/73687

www.igi-global.com

701 East Chocolate Avenue, Hershey, PA 17033, USA
Tel: 717-533-8845 x100 • Fax: 717-533-8661
E-Mail: cust@igi-global.com • www.igi-global.com

Table of Contents

Preface vi

Acknowledgment xii

Chapter 1
Synthesis, Characterization, and Catalytic Application of 2D Mo(W) Dichalcogenides Nanosheets 1

Chapter 2
The Fundamental Research and Application Progress of 2D Layer Mo(W) S2-Based Catalyst 31

Chapter 3
3D Catalysts of Mo(W) Carbide, Nitride, Oxide, Phosphide, and Boride 53

Chapter 4
Low-Dimensional Molybdenum-Based Catalytic Materials from Theoretical Perspectives 100

Chapter 5
Summary and Perspectives 129

Related Readings 137

About the Authors 158

Index 160

Preface

The discovery of graphene has attracted tremendous research interest in two-dimensional (2D) layered materials characterized with strong covalent bonding in intralayer and weak van der Waals interaction between interlayers. Among them, Mo and W dichalcogens (e.g. WS_2, MoS_2, $MoSe_2$, $MoTe_2$) have been deeply and extensively explored due to their special properties and easy fabrication. These 2D materials with reduced dimensionality in vertical direction exhibit unique catalytic properties. For example, the band structures can significantly be altered when these materials are reduced from bulk to the single-layer limit, giving rise to great opportunities to fabricate catalysts with excellent performance. Besides, these 2D materials can be semiconductors, metals, and superconductors, and they represent a versatile and ideal system for exploring catalysis at the limit of atomic scale, and have the potential to open up exciting new opportunities beyond the reach of existing materials and enable advances across diverse disciplines, such as electronics, photonics, energy and catalysis.

Typical layer of Mo or W dichalcogenide features the "sandwich" layer, which consist typically of one plane of hexagonally packed metal atoms sandwiched by two planes of chalcogenide atoms. The sandwich layers are vertically stacked and loosely bonded by weak van der Waals forces to form the 3D bulk material. This high anisotropy induces the structure-sensitive catalytic phenomena. However, a material with fixed structures may not exhibit versatile applications. With the unique anisotropy, the physical and chemical properties of 2D layered Mo(W) dichalcogenides can be tuned easily through different strategies such as reducing dimensions, intercalation, and alloying and new catalytic properties can be achieved. For example, through the intercalation of guest alkali ions, the carrier densities of 2D layer Mo(W) dichalcogenides can be increased by multiple orders of magnitude, companied by a transformation of the Mo(W) dichalcogenides from semiconductor 2H phase to 1T metal transition. This results in the large improvement of hydrogen evolution reaction (HER).

Another interesting feature is that Mo(W) dichalcogenide 2D layers can constitute lateral or vertical heterostructures. Vertical heterostructures which are held together by van der Waals forces can be fabricated by almost arbitrary combinations using relative simple synthesizing techniques. However, lateral heterostructures with two materials linked by covalent bonds are more difficult to manufacture. These heterostructures of 2D Mo(W) dichalcogenide layers create superlattice and complexity with totally new functions for electric- and photo-catalysis. Therefore, Mo(W) dichalcogenide 2D heterostructure prepared by lateral or vertical growth or alloying provide almost unlimited opportunities to synthesize excellent catalysts with tunable properties. Besides dichalcogenides/dichalcogenides heterostructures, the flexible van der Waals interaction can also allow the creation of diverse heterostructures between 2D Mo(W) dichalcogenides and other atomic layered materials, including graphene, carbon nitride and boron nitride etc. The formation of such versatile heterostructures can enable the design of entirely new "interbedded" catalysts.

To utilize the full potential of 2D Mo or W dichalcogenide materials, it is necessary to develop strategies for synthesizing these materials with a well-defined dimension, chemical composition, and heterostructure interface. The bottom-up approach of chemical vapour deposition (CVD) allows the growth of relatively high quality 2D layers on supported surfaces with a well-defiined lateral size and layer thickness. Hydrothermal or solvothermal synthesis is also promising but with only little investigation. Among the top-down approaches, the mechanical exfoliation of bulk crystals have produced large areas of 2D Mo(W) dichalcogenides with variable thicknesses down to a single lattice unit, and has been utilized for initial fundamental studies. It is however limited by the very low throughput. The chemical and solvent exfoliation approach can allow for the production of a relatively large quantity of 2D materials. This are now the only possible method for preparing the industrial 2D catalysts. But the resulting materials may be contaminated with impurity doping or have a poor control of the dimension size and thickness. Thus the synthesizing approaches still need to be developed.

Mo(W)S_2-based layer catalysts in face has long been used in petroleum and chemical industry before the spring up of the graphene. They are of critical importance in the refining processes, such as hydrotreatment, hydrocracking and hydrogenation. The hydrotreatment is seen as an inheritance of coal technologies developed in Germany at the beginning of the twentieth century. In 1924 researchers of BASF found that transition metal sulfides were efficient catalysts for coal hydro-liquefaction and screened out molybdenum sulfide used in these processes. After the Second World War, instead of the coal conversion, petroleum refining got fast development. Molybdenum sulfides promoted by cobalt and supported on alumina were firstly used in the USA as the catalysts for the hydrotreating processes which simultaneous removed sulfur, nitrogen and metal impurities in oil streams. Since that time, new combinations of supported sulfides (such as NiMo and NiW) had also been synthesized for more specific applications, e.g. hydrogenation and hydrocracking.

The oil crises in 1973 and 1979 stimulated a surge of researches related to these catalysts and processes in both academic laboratories and refining industries with the objective of developing more active and selective catalysts. During 1980s and 90s, accompanying with the advance in analytic and characterized technologies, important discoveries were reported in the literature. For example, using Mössbauer spectroscopy, the interaction of Co and Mo was elucidated as the Co-Mo-S active phase responsible for the promotion in hydrodesulfurization (HDS) catalytic activity. It was proposed that these Co-Mo-S (and also Ni-Mo-S) structures are small MoS_2-like nanocrystals with the promoter atoms located at the edges of the MoS_2 layers. Further investigation suggested that Co atoms are located in the same plane as Mo, but that their local coordination is different.

However for a long-time it is difficult to address the issue of the detailed edge structure of unpromoted and promoted MoS_2 2D layer, as atomic-resolved structures could not be detected. Entering 2000s, with the fast advance in characterization at atomic level, some breakthroughs were achieved. By scanning tunneling microscopy (STM), the real-space structures of MoS_2 and Co(Ni) promoted MoS_2 nanoclusters grown on flat model substrates were imaged. It was possible, for the first time, to reveal the equilibrium morphology of these 2D nanoclusters. And the brim sites, which is located near the edge, was firstly discovered by the STM experiment, and DFT calculation suggested it as the active centers for hydrogenation reaction due to the metallic property. By high-angle annular dark field scanning tunneling electron microscopy (HAADFSTEM), additional information on the morphology of $Mo(W)S_2$ based nanostructures have also been obtained. Furthermore, with the combination of these new technologies, it is now possible to elucidate the detailed $Co(Ni)Mo(W)S_2$ structure, such as the layer edges, the sulfur coverage and the sulfur vacancies or defects, which are considered to be responsible for the catalytic properties.

Based on the rich understanding about the relation of structure and activity, now designing HDS catalysts for controlled structure and functionalities are becoming a trend. It appears that now science is catching up with technology. Before long, due to the lack of adequate characterized tools, a detailed fundamental understanding of $Mo(W)S_2$ based catalysts and their reaction mechanisms was unavailable. At that time, catalyst development was to a large extent based on the "trial-and-error" experimentation.

Although the activity and selectivity of $Mo(W)S_2$ based catalyst have been largely improved, development of better catalysts for hydrotreatment is still urgent. Environmental regulation in many countries requires the sulfur content in transport fuels down to about 10 ppm with the aim of reducing engine's harmful emissions and improving air quality. And at the same times, the quality of transport fuels must be guaranteed. The ultra-desulfurization of gasoline streams needs to depress the hydrogenation of olefins to keep the octane number and decrease the hydrogen consumption. Meanwhile the production of ultra-low sulfur diesel (ULSD) necessitate some hydrogenation to aromatics increasing the cetane number. Thus the improvement of the hydrogenation selectivity will play a critical role in future hydrotreatment processing.

The application of $Mo(W)S_2$ based catalyst are also been extended to other ranges, such as CO_2 electric reduction, electro- and photo- catalytic water-splitting, water-gas shift, and hydrotreating of bio-fuel. The vast emissions of CO_2 through combustion of carbonaceous fuels, such as coal, oil, natural gas and wood, have evoked social concerns due to the greenhouse gas effect. In the past few years, CO_2 chemistry has become a very dynamic area of research, hoping the use of emitted CO_2 as a potential alternative and economical C1 feedstock. The recent advances in the electric reduction of CO_2 to CO in ion liquid using MoS_2 layer flakes have demonstrated the potential of Mo based catalyst in CO_2 conversion. The hydrogen evolution reaction (HER) via an electrocatalytic water-splitting method is considered a sustainable approach for hydrogen production if the electricity is from renewable. The key problem now is seeking highly active electrocatalysts that can decrease the overpotential (η) and promote the HER performance. Pt is now the most efficient electrocatalyst for the HER, but its low abundance and high cost prevent large scale applications. Thus, the development of non-noble metal catalysts with low cost and high catalytic activity has attracted great research interest. In this field, a series of 3d transition metals, such as WN, WO_2, WS_2, MoO_2, MoS_2, MoB, MoP, $MoSe_2$ and Mo_2C, have been exploited as potential substitutes for Pt-based catalysts.

Since Levy and Boudart reported in 1973 that tungsten carbides behaved similar to platinum for some types of reactions, a surge of research in these types of materials have occurred. It was evidenced that Mo(W) carbides in bulk or supported form, as well as promoted or not, are active for a lot of reactions that are usually catalyzed by noble metals. These carbides (e.g. Mo_2C and WC) are formed by the incorporation of carbon atoms into the metal lattice and display intriguing catalytic properties. Numerous studies have shown that the carbides of molybdenum and tungsten exhibit catalytic activities for chemical reactions, such as ammonia synthesis, carbon monoxide hydrogenation, water gas shift, methane reforming, and HDS. The turnover rates over carbide catalysts for such reactions under reducing environments were equal to or greater than those of noble metals (e.g. Pt, Pd and Ru) supported on oxide, and these carbide materials were deemed to be cheaper replacements for noble metal catalysts. Other Mo(W) based materials, such as nitride, phosphide, boride or oxide, usually possess some special properties which can be used in a lot of chemistry reactions to synthesize chemicals or improve the quality of fuels

Research on sustainable energy has become hotspot in recent years due to the continued depletion of fossil energy resources and the increased concentration of CO_2 in the atmosphere which is responsible for global warming. Biomass, as a renewable carbon source, can be converted into liquid fuels after fast pyrolysis or hydrothermal liquefaction. The process of producing bio-derived fuels and chemicals can be integrated into the petroleum refining and chemical industry. However, these bio-derived oils contain many oxygen-containing compounds, including phenols, furans, aldehydes, alcohols and esters, which lead to the high oxygen content, low heating value, high viscosity and instability. To solve these problems, the oxygen content in these bio-fuels has to be lowered by a proper upgrading process. Hydrodeoxygenation (HDO) is a recommend technology for the selective hydrogenolysis of C–O bond to remove oxygen at high temperature and hydrogen pressure, and the oxygen-removal efficiency largely depends on the selectivity and activity of catalysts. Currently, high catalyst cost, short catalyst lifetime, and lack of effective regeneration methods are hampering the development of this otherwise attractive renewable hydrocarbon technology. Meanwhile sustainable production of chemicals and fuels via the deoxygenation (DO) of plant oil, animal fat or waste oils has become a prominent research interest. Mo(W) based sulfides, carbides and nitrides have shown some potentials in this regard. However, the diversity of oxygen-containing functionalities in compounds of biomass-derived fuels necessitates robust and multifunctional catalysis for the successful upgrading if applicable. Further improvement of Mo(W) based catalysts for the HDO of bio- derived oil is still challenging.

In this book, we start with a review of the exciting advances in synthesis and catalytic application of 2D layer Mo(W) dichalcogenides. Then we revisit the traditional Mo(W)S_2 based catalysts with a focus on the newly progress on the insight of their atomic active structures and reaction mechanism, and show how the Mo(W)S_2 based catalysts have been improved based on the renewed knowledge and novel synthesizing technique. We further discuss the extension of these catalysts to novel application in petroleum and chemical industry. The novel synthesis and application of Mo(W) based materials other than the dichalcogenides, such as carbide, nitride, phosphide, are collected in a separated chapter. The followed chapter brings in theoretical perspectives on the properties of these materials and related reaction mechanisms. And we summary and project the future in the last chapter. We hope that the rich knowledge can bring the readers entering the colorful world of Mo(W) based catalysis.

Acknowledgment

We would like to acknowledge Colleen Moore and Jan Travers at IGI Global. We could not have achieved this book smoothly without their kind help. We would like to express our gratitude to the many people who assisted in the editing, proofreading and design. The authors are grateful for the National Natural Science Foundation of China (21473231). Above all Dr. Ge want to thank his parents, wife and children. Last and not least, we beg forgiveness of all those who have helped and whose names we have failed to mention. Many thanks

Chapter 1
Synthesis, Characterization, and Catalytic Application of 2D Mo(W) Dichalcogenides Nanosheets

Since Novoselov et al. (2004) discovered the novel properties of monolayer graphene, there emerges a wave for researching 2D materials. By tuning the number of atomic layers of these materials unprecedented properties can be achieved. With the reduced dimensionality as well as quantum confinement effect, 2D materials can exhibit unique catalytic properties distinct from their 3D bulk counterparts. Now 2D material at the limit of single-layer thickness has been explored for numerous catalysis reactions, opening up new technological opportunities beyond the reach of existing materials.

2D Mo(W) dichalcogenides (e.g. MoS_2, WS_2, WSe_2 WTe_2) represent a large system of 2D materials which have the unique "sandwich" structure consisting of a transition metal layer between two chalcogen layers. They are characterized by the weak non-covalent bonding between the "sandwich" layers and strong covalent bonding in plane. Owing to this special structural

DOI: 10.4018/978-1-5225-2274-4.ch001

property, bulk Mo(W) dichalcogenides can be easily exfoliated into single- or few-layered 2D materials, which can be further tuned or assembled to heterostructures (Figure 1). Scalable production of Mo(W) dichalcogenides have fast progressed recently, affording their potential application as effective catalysts in petroleum and chemical industry (Li et al. 2014a).

In this chapter, we firstly introduce synthesizing methods of 2D Mo(W) dichalcogenides nanosheets in section 1. Then we present the assembling of heterostructures between Mo(W) dichlcogenides and between dichalcogenides with other 2D nanosheets in section 2, In 3 section, we discuss the tuning of Mo(W) dichalcogenide layers to achieve special catalytic functionality. And then we illustrate the characterizations and the reaction mechanisms of 2D Mo(W) dichalcogenides materials in section 4, Chapter 2. In last section, we show the novel application researches utilizing these 2D Mo(W) materials.

Figure 1. (a) Schematic tuning 2D Mo(W) dichalcogenides properties by guest ion intercalation and reducing dimension along the z direction or xy directions. (b) Tuning 2D Mo(W) dichalcogenides properties by constructing heterostructures and alloying.

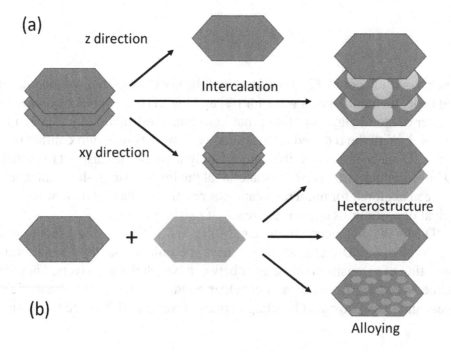

1. PREPARATION OF Mo(W)-BASED 2D NANOSHEETS

The controlled preparation of the Mo(W) dichalcogenides 2D layer materials is of critical importance in order to carry out their catalytic applications. The structural and electronic properties of these 2D nanosheets, at large extent, dominates the activity, selectivity and durability of catalytic reactions. Due to the weak interaction between layers, Mo(W) dichlcogenides down to single or several layers can be prepared more easily for versatile applications. The preparing routes to 2D layer Mo(W) dichlcogenides can be roughly classified as top-down (e.g., adhesive tape exfoliation, ultrasonic assisted exfoliation, solvent or surfactant or polymer assisted exfoliation, and chemical exfoliation via lithium intercalation) and bottom up (e.g., direct wet chemical synthesis and chemical vapor deposition) approaches (Frindt 1966, Nicolosi et al. 2013, Zeng et al. 2012, Xiao et al. 2012 and Mak et al. 2012). Both approaches have attracts interests from various disciplines, such as electronics, photonics, materials, energy, environment and catalysis.

1.1 Exfoliated Method

Among the top-down exfoliation approaches, mechanical exfoliation of bulk crystals, intrigued by the preparation of graphene, can fabricate the ultraclean 2D Mo(W)-based nanosheets even down to single layer. It is however limited by the low productivity. Conventionally, bulky Mo(W) dichalcogenide materials was exfoliated by intercalation of lithium metal followed by reaction with water (Joensen et al. 1986). The intercalated lithium metal reacts violently with water producing hydrogen gas causing the layers to separate, thus resulting in a colloidal suspension in water (Ramakrishna Matte et al. 2010). Both theoretical and experimental evidence suggests that the intercalation is accompanied by a change in the crystal symmetry from 2H to 1T (Heising et al. 1999). And harsh preparing conditions led to defects and the introduction of phase impurity in the obtained 2D Mo(W) dichalcogenides nanosheets, which alter the electronic properties of the product. Meanwhile, the reagents used in such processes, especially n-butyl lithium, are highly flammable and explosive, posing health hazards and complicating the upscale of such processes to industrial quantities. For avoiding the use of dangerous chemicals, Zeng et al. (2011) fabricated single-layer 2D MoS_2 and WS_2 nano-materials by electrochemical intercalation of Li^+ into the interlayer using a cell system. The lithium plays dual roles. First, with the insertion of Li^+ ions spaces, the interlayer distance of the bulk materials is expanded, weakening the van der Waals interactions between the layers. Second, metallic Li (after insertion

Li + ions are reduced by electrons during the discharge process) reacts with water to form Li(OH) and H_2 gas which pushes the layers further apart. After providing sufficient agitation by ultrasonication, isolated 2D nanosheets was obtained. Compared to the traditional lithium intercalation, electrochemical method can be easily conducted at room temperature with shorter time. And the lithiation process can be monitored and finely controlled in the cell test system. Unfortunately, this method results in low yields (Wang et al. 2013a and 2015). Thus chemical intercalation of ammonium is deemed a promising alternative. The ammoniated $Mo(W)S_2$ bulky crystals are expanded due to that the intercalated NH_3/$NH4^+$ brings in solvent molecules into the interlayer. And ultrasonication and NH_3 release assists the exfoliation (Jeffery et al. 2014).

Liquid phase exfoliation approaches allow for production of a relatively large quantity of 2D Mo(W) based materials. Furthermore they have additional advantages, such as the ease of preparation of nanocomposites, layered hybrids, and fabrication of thin films. The bulky crystallites are usually ultrasonicated in certain stabilizing liquids using appropriate solvents, surfactants or polymers (Zhou et al. 2011 and Cunningham et al. 2012). The produced nanosheets are stabilized against aggregation via the interaction of the liquid and/or guest compounds. So far, liquid phase exfoliation techniques have shown to be the highest yielding technique when starting from a bulk material (Halim et al. 2013 and Nicollosi et al 2013). These methods are favorable for industrial application, as they do not involve any chemical reactions and are not air-sensitive.

As for solvent exfoliation methods, the selection of solvent plays an important part since physical properties, such as boiling points, surface tension and energy, as well as solubility parameters affect the yield and quality of resulting 2D material (Zhou et al. 2011). Coleman et al. (2011) compiled a list of solvents and evaluated their suitability for the sonication assisted exfoliation of 2D materials. The most effective solvent was found to be N-methylpyrrolidone (NMP). However, this solvent is difficult to be removed due to a relatively high boiling point. Therefore, residual NMP are often found on the resulting 2D materials, even when post cleaning process are used. Therefore, Nguyen et al. (2015) in their grinding-assisted liquid phase exfoliation employed a two-solvent step approach for the fabrication of 2D layer MoS_2. They implemented a separate solvent during the grinding phase and ethanol during exfoliation. The grinding solvent was found to play a critical role, determining exfoliation yield, flake dimensions, and morphology. Acetonitrile was identified as a promising alternative to NMP as the grinding solvent due to satisfactory exfoliation yields and the complete removal from the MoS_2 surface after grinding. And the obtained nanosheets were smaller when compared to NMP ground MoS_2.

Recently it was demonstrated that liquid exfoliation can also be achieved by exposing bulk crystals to high shear rates using either rotor-stator high shear mixers or simple kitchen blenders (Varrla et al. 2015 and Chen et al. 2012). Sodium cholate is commonly used as the surfactant to stabilize the liquid-exfoliated layered materials. Incorporating grinding into the exfoliation procedure, Yao et al. (2013) obtained relatively higher yields of 2D layer Mo dicalcogenides, up to 26.7 mg/mL. However the challenge of controlling the lateral growth without expansion in the c-axis direction is still unsolved.

1.2 CVD Method

The bottom-up chemical vapor deposition (CVD) process can allow the growth of relatively high quality 2D crystals on surfaces with a well-controlled lateral size and layer thickness (Duan et al. 2015). In most of the reported studies, 2D layer Mo(W) dicalcogenides are grown by thermal CVD process, where the chemical vapor is generated by thermal evaporation of a solid source and an inert gas (e.g., argon) is used as the vapor phase carrier (Figure 2). In some cases (Lee et. al. 2012), the presence of hydrogen can promote the formation of atomically thin nanosheets (e.g., $MoSe_2$).

CVD processes producing MoS_2 nanosheets can use different precursors with slight different procedures, such as:

1. 2. Coating $(NH_4)_2MoS_4$ on a substrate followed by annealing in sulfur vapor;
2. Vaporizing MoO_3 (or $MoCl_5$) and S for co-deposition on a substrate;
3. Sulfur vapor treating Mo thin films on a substrate;
4. Sulfur vapor treatment of MoO_2 thin films on a substrate;
5. Vaporizing MoS_2 at high temperature and vapor transfer deposition on a substrate at lower temperature.

These methods are general and can be readily extended to the growth of other dichalcogenides, e.g. WS_2, $MoSe_2$, and WSe_2. In some cases, seeds are used to initiate the growth of atomically thick layer crystal. For example, perylene-3,4,9,10-tetracarboxylic acid tetrapotassium salt (PTAS) is used as an effective seed material for growing MoS_2 and WS_2 monolayers (Liu et al. 2012 and Duan et al. 2015).

Figure 2. Reactor-conditions-dependent WSe$_2$ growth with varying temperature and flow rate of argon. Optical microscope images of WSe$_2$ samples grown at different temperature under designed flow rate for 20 min: (a) 750 °C, 100 sccm (inset: SEM image of same size sample, indicating high density but small nucleations (~300 to 500 nm); scale bar, 5um); (b) 765 °C, 100 sccm; (c) 780 °C, 100 sccm; (d) 795 °C, 100 sccm; (e) 750 °C, 150 sccm; (f) 765 °C, 150 sccm; (g) 780 °C, 150 sccm; (h) 795 °C, 150 sccm; (i) 750 °C, 200 sccm; (j) 765 °C, 200 sccm; (k) 780 °C, 200 sccm; (l) 795 °C, 200 sccm. All of the scale bars are 20 µm.
Source: Zhou et al., 2015a.

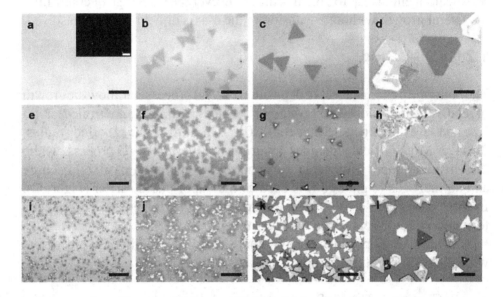

In general, when the super-saturation degree of a chemical precursor in a CVD process exceeds a certain threshold, the growth occurs either on the edge of the existing layer to extend the lateral 2D growth, or on the top of the existing layer to initiate the vertical 3D growth equivalent to the formation of new nuclei on the existing template. To grow the nanosheets of Mo(W) dichalcogenides with controlled lateral and vertical dimensions, the critical requirement is to independently control the nucleation and growth steps in 2D and/or 3D dimensions. It has been well established that the nucleation probability is proportional to the super-saturation of vapor precursor and inversely proportional to the substrate temperature. To exert exquisite control of the growth process, it is necessary to have a fine control over the growth temperature, the flow rate, super-saturation and spatial distribution uniformity of the vapor phase reactants. However, using the conventional thermally evaporated source, it is usually difficult to precisely control the vapor partial pressure and spatial distribution of the chemical vapor precursors. Recently, metal organic CVD (MOCVD) process has been suggested for the improved control but in a more sophisticate way (Kang et al. 2015). Until now, it remains a significant challenge to control of the crystal domain size, domain density and the layer thickness precisely.

The thermal CVD process has been proven to be a rather flexible approach for successful growth of diverse 2D Mo(W) dichalcogenides nanosheets with the grain size ranging from1 from10 micrometers. It is important to note that the continuous atomic thin films resulting from most of the CVD or MOCVD systems to date are typically polycrystalline materials with defects and grain boundaries (van der Zande, et al. 2013 and Najmaei et al. 2013). This offer exciting opportunities for efficient catalysis which needs sufficient edge surface or defects to carry on the reaction.

1.3 Wet Chemical Method

Bottom-up wet chemical methods offer a potentially powerful routes for synthesizing 2D Mo(W) dichalcogenides where nanosheets form in high yield directly in solution. With solvothermal method using $(NH_4)_6Mo_7O_{24} \cdot 4H_2O$ and thiourea pecursors, Xie et al. 2013 synthesized MoS_2 2D nanosheets. Due to the simplicity and versatile, this method is fitful to synthesize catalysts in industry scale. However, this approach have not been sufficient investigated and developed.

In summary, the solvent or surfactant assisted exfoliation and wet chemical methods can produce 2D nanosheets in larger quantities. This provide the opportunities for synthesizing 2D layer Mo(W) dichalcogenides catalytic materials for the potential application in petroleum and chemical industry. On the other hand, CVD method synthesizes 2D nanosheets with limits of the low productivity, which is more fitful for fundamental research or synthesizing catalysts for fine chemicals.

2. HYBRIDIZATION OF Mo(W) DICHALCOGENIDES 2D NANOSHEETS

Different 2D layer Mo(W) dichalcogenides materials may be vertically overlapped to create heterostructures or superlattices to produce complexity functions. A distinct feature of these 2D materials is the van der Waals interactions between neighboring layers. This allows flexible integration of different 2D materials without the limitation of lattice matching. Thus vast possibilities are opened up for nearly arbitrarily combination (Geim and Grigorieva 2013).

Colloidal dispersions of 2D nanosheets of MoS_2/WS_2 have been fabricated either by layer-by-layer deposition of Langmuir-Blodgett technique or by self-organizational assembly of a mixture of suspensions of different 2D Mo(W) dichalcogenides layers (Osada and Sadaki 2012).

CVD is proven an effective method for growth of well-defined heterostructures with designated chemical compositions and/or electronic structures in the lateral or vertical direction. In principle, a lateral heterostructure can be produced by successive growth of a second material (e.g., $MoSe_2$, WSe_2) at the edge of the first layer material (e.g., MoS_2, WS_2). Since the lattice mismatch between these materials is <5%, heterostructures with nearly perfect lattice structures could be produced by the lateral growth process (Duan et al. 2014). On the other hand, the vertical heterostructures can be synthesized using a layer-by-layer CVD process.

Alloying semiconductors producing different band gaps represents a common strategy for the band gap engineering for electric or photo catalysis. Considering the similarities in the atomic structure of many 2D Mo(W) dichalcogenides, it is possible to create a mixed alloy system with tunable band gaps depending on the compositions. Theoretical studies suggested that mixed $Mo(S, Se, Te)_2$ heterostructurs are thermodynamically stable. And compositions can be tuned continuously between the constituent limits (Komsa and Krasheninnikov 2012, Mann et. al. 2014 and Li et al. 2015). Using a spatial temperature gradient for the composition selection, Shaw et al. (2014) synthesized the composition tunable MoS_{2x}-$Se_{2(1-x)}$ nanosheets. Elemental analysis at different growth temperatures shows clearly modulation of composition from $x = 0$ until to $x = 1$. And the band gap modulating within the alloyed nanosheets was demonstrated by the photoluminescence experiments. Terrones et al. (2013) predicted that MoS_2-WSe_2 heterostructures can exhibit electronic properties entirely different from their constituent layers; and significantly decreased bandgap energy may be achieved. This has been realized experimentally by Fang et al (2014) recently. The bandgap engineering has the large potential to control the catalytic reaction at atom levels.

Apart from the heterostructures between different Mo(W) dichalcogenides, the van der Waals interaction can allow also the flexible creation of diverse heterostructures between 2D layer Mo(W) dichalcogenides and other layered materials, such as graphene and boron nitride. A variety of such heterostructures have been synthesized with precise interfaces and atomic arrangements based on MoS_2, WSe_2, hBN, and graphene. The formation of such heterostructures enable the design of entirely new catalysts with unique attributes unavailable with traditional methods.

For example, Li et al. (2011) synthesized the MoS_2/RGO heterostructures using solvothermal reaction of $(NH_4)_2MoS_4$ and hydrazine in an N,N-dimethylformamide (DMF) solution of grapheme oxide (GO). Chemical interactions afford the selective growth of highly dispersed MoS_2 nanoparticles on GO. The obtained MoS_2/RGO catalyst exhibited excellent hydrogen evolution reaction (HER) activity with a very low overpotential of ~0.1 V and a small Tafel slope of 41 mV/decade. It is well established that the unsaturated S sites are responsible for the catalytic reactions. The small size MoS_2 shows an abundance of accessible edges that could serve as active catalytic sites for the HER. Electrical coupling of MoS_2 to the underlying graphene sheets provides rapid electron transport from the less-conducting MoS_2 nanoparticles to the electrodes. On the contrast, in the absence of GO, the exact same synthesis method produced MoS_2 coalesced into 3D-like particles of various sizes (Figure 3). Similar synthesized methods were tried to prepare the amorphous MoS_x on $g-C_3N_4$ which is shown efficient for photo-catalytic H_2O splitting (Yu et al. 2016).

Lateral heterostructures of $WS_2/WO_3 \cdot H_2O$ have been successfully constructed by exfoliating monolayer or few-layer WS_2 from bulk WS_2 with the assistance of supercritical CO_2 and then partially converting them into tung-

Figure 3. Synthesis of MoS_2 in solution with and without graphene sheets. (A) Schematic solvothermal synthesis with GO sheets to afford the MoS_2/RGO hybrid. (B) SEM and (inset) TEM images of the MoS2/RGO hybrid. (C) Schematic solvothermal synthesis without any GO sheets, resulting in large, free MoS_2 particles. (D) SEM and (inset) TEM images of the free particles.
Source: Reprinted with permission from Li et al. (2011). Copyright of American Chemical Society.

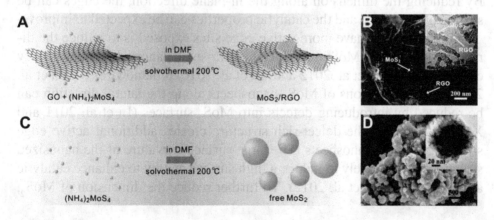

sten oxide monohydrate ($WO_3 \cdot H_2O$) through oxidation (Zhou et al. 2015b). The sc-CO_2 permeating between WS_2 layers expands the interlayer distance, leading to single and few-layer WS_2 nanosheets, due to more active edges exposed, oxidation of WS_2 are beneficial. The formed heterostructures allows for separation of long-lived electron-hole pairs, thus resulting in enhanced photocatalytic activity toward the degradation of methyl orange and higher photocurrent under visible-light irradiation.

3. TUNING OF Mo(W) DICHACOGENIDES NANOSHEETS

It becomes increasingly important to understand how the proprieties of 2D material can be tuned and how the tunable properties can be utilized in catalysis. For this regard, we here focus on the tuning properties of 2D layer Mo(W) dichacogenides to improve their catalysis efficiency.

3.1 Dimension Tuning

Nano-Mo(W) dichacogenides have been prepared by different approaches, including the gas-solid synthesis method, precursor thermal decomposition method, hydrothermal method, and electrochemical deposition method. And obtained Mo(W) dichacogenides materials can form different morphologies such as nanorods, nanoplates, nanoribbons, nanotubes, globular shapes, flower-like structures and fullerenelike nanoparticles etc. (Kong et al. 2013 and Dunne et al. 2015). The 2D Mo(W)S_2 "sandwich" layer are their basic building block whose edges usually have unsaturated coordination and dangling bonds, offering the interesting and important active sits for reaction molecules (Jaramillo et al. 2007, Huang et al. 2013 and Zeng et al. 2014). By reducing the dimension along the in-plane direction, the edges can be sufficiently exposed and the catalytic properties can be expected to improve.

Another way to have more active edge sites exposed is to reduce the dimension of the bulk MoS_2 material into ultra-small nanoparticles (Afanasiev et al.2012, Yoosuk et al. 2012 and Yu et al. 2016) or nanowires (Chen et al. 2011). The dimensions of MoS_2 nanosheets along the lateral direction can be reduced by introducing defects into MoS_2 surfaces (Li et al. 2011 and Xie et al. 2013). The defect-rich structure creates additional active edge sites in the MoS_2 nanosheets. The large surface curvature of the nanosized structures is also likely to induce a high surface energy to enhance catalytic activity (Kibsgaard et al. 2012). To further reduce the dimension of MoS_2, zero-dimension molecular MoS_2 edge site mimic has been successfully prepared with very high HER activity, demonstrating the importance of edge site exposure (Karunadasa et al. 2012).

3.2 Intercalation Tuning

Studies have revealed the strong correlation between the electronic structure and the catalytic activity through theoretical simulations as well as experiments. Proper electronic structures of the active sites can be designed to create suitable chemical bonding with the reactants (not too weak and not too strong), which ensures both a good electron transfer between the catalyst and the reagent and a facile products-releasing process (Hinnemann et al. 2005). Therefore, the tunable electronic structure through electrochemical or chemical intercalation makes 2D Mo(W) dichalcogenides very attractive candidates for catalysis optimization. Electrochemical intercalation can effectively shift the chemical potentials of 2D Mo(W) dichalcogenides materials to the optimized position for efficient catalysis. Sometimes the intercalation process introduces a phase transition of the host matrix, which may change catalytic activity and selectivity. In addition, the intercalated guest atoms and the matrix material may have charge transfers between each other, which increases the carrier density and thus improves the conductivity of the catalyst (Wang et al. 2015).

A successful example is lithium electrochemical tuning of 2H MoS_2 for enhanced HER activity (Wang et al. 2013a and 2013b); as shown in Figure 4, the edge-terminated MoS_2 nanofilm and a piece of Li foil were made into a battery cell to perform Li intercalation. A galvanostatic discharge provides the MoS_2 electronic structure change such as the reduction of Mo oxidation state and the MoS_2 transition from 2H to 1T phase (Figure 4c). The intercalated Li transfers excess charge carriers to MoS_2, increasing the electronic energy of the whole system, and thus inducing the structure phase transition for a more stable octahedral coordination. This electronic structure change improves the HER catalytic activity significantly. The chemical potential of Li intercalated MoS_2 was continuously shifted to 1.1 V vs. Li+/Li. Consequently, the HER performance is continuously improved, with the Tafel slopes improved from 123 mV per decade all the way to 44 mV per decade. It noted that the 1T phase MoS_2 is still stable even though the Li inside reacts with air and water (Voiry et al. 2013a).

Another way to tune the electronic structure of MoS_2 and WS_2 nanosheets for improved HER activity is chemical intercalation. WS_2 nanosheets have been successfully prepared with 1T phase by a chemical intercalation and exfoliation process (Voiry et al. 2013b). It is evidenced that the hexagonal

Figure 4. (a) Schematic structure of MoS_2. (b) Schematic of the edge-terminated MoS_2 lithiation process. (c) Galvanostatic discharge curve with schematics of charge transfer and phase transition.
Source: Wang et al. 2013b.

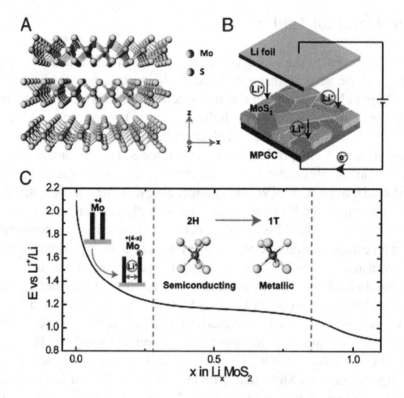

lattice 2H WS_2 undergoes a phase transition to form a distorted 1T structure, leading to the significant improvement for HER performance. It was revealed by the theoretical simulation that showed the strains in the 1T nanosheets help to lower the free energy of reaction.

The solution of n-butyl lithium chemically can intercalate a large amount of Li atoms into the MoS_2. This shifts the MoS_2 to low chemical potential, exfoliates the nanosheets into single or few layers and induces the transition from 2H to 1T phase. As a results, the HER performance was tremendously improved with a Tafel slope of 43 mV per decade (Lukowski et al. 2013 and 2014). The techniques of intercalation tuning of 2D Mo(W) dichalcogenides have constituted a vast pool for improvement of catalysts. For example, the electrochemical tuning method has also been demonstrated to be effective to enhance electrocatalysis of other 2D materials such as lithium transition metal oxides (Lu et al. 2014)

4. THE CRITICAL CHARACTERIZATION AND CATALYTIC MECHANISMS OF 2D Mo(W) DICHALCOGENIDES NANOSHEETS

Depending on the atomic stacking configurations, Mo(W) dichalcogenides can form mainly three phases, namely, trigonal prismatic (2H) phase, octahedral (1T) phase and (3R) phase (Figure 5). In thermodynamically stable 2H-MoS_2 phase, each Mo atom is prismatically coordinated to six surrounding S atoms, whereas for the 1T-MoS_2 phase, six S atoms form a distorted octahedron around one Mo atom, leading a metastable phase. Interestingly, two phases can be transformed to the other via atomic gliding. 1T-MoS_2 due to the thermodynamically instability gradually converts back to 2H-MoS_2 at room temperature (Sandoval et al. 1991). On the other hand, 1T-MoS_2 can also be converted from 2H-MoS_2 by intercalating Li or K (Wang et al. 2014).

Figure 5. Structure of MoS_2. They can crystallize in 1T (tetragonal), 2H (hexagonal), or 3R (rhombohedral) symmetry. The figure also shows the hexagonal Brillouin zone with the high symmetry k points.
Source: Reprinted with permission from Heine (2015). Copyright of American Chemical Society.

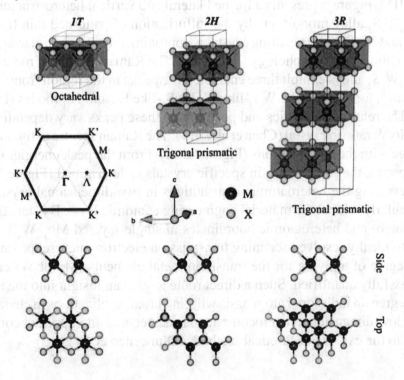

2H-MoS$_2$ is semiconducting, while 1T-MoS$_2$ is metallic (Lin et al. 2014). Due to their apparent catalytic properties, it is necessary to develop simple methods that can easily identify 1T and 2H phases. Raman spectroscopy has been developed for this object. It is observed that both 1T-MoS$_2$ and 2H-MoS$_2$ show A$_{1g}$ and E$_{2g}$ vibrational modes at ~380 and ~410 cm^{-1}, respectively. Owing to symmetry differences, 1T-MoS$_2$ presents several additional Raman vibrational modes at~160 cm^{-1}, ~230 cm^{-1} and~330 cm^{-1}, which do not appear for 2H-MoS$_2$ (Lv et al.2015). However, accurate quantitative analysis of the phase composition is not easy only from the Raman spectra due to weak Raman response of these vibration modes. X-ray photoelectron spectroscopy (XPS) is an efficient technique for quantifying the 1T and 2H phases. The XPS spectra in the Mo 3d region usually can be decomposed two peaks near 230 and 232 eV binding energies, corresponding to photoelectrons from the Mo^{4+} 3d$_{5/2}$ and Mo^{4+} 3d$_{3/2}$ states, respectively. The Mo 3d region of 1T-MoS$_2$ presents peaks at ~228.1 and ~231.1 eV, which are slightly lower than those for 2H-MoS$_2$ (~229.5 and ~232.0 eV). Therefore, by deconvoluting the peaks of Mo^{4+} 3d$_{5/2}$ and Mo^{4+} 3d$_{3/2}$, the relative concentrations of the 1T and 2H phases can be obtained.

Heterostructures of mixed Mo and W dichalcogenides can be characterized sufficiently by PL optical microscopy and Raman etc. (Kobayashi et al. 2015). Figure 6 presents a formed lateral and vertical heterostructures of Mo$_{1-x}$W$_x$S$_2$ alloy monolayers by the sulfurization of patterned thin films of WO$_3$ and MoO$_3$. The resulting crystal is approximately 20 μm in size and has a six-pointed star morphology (Figure 6a). The Raman spectra of monolayer Mo$_{1-x}$W$_x$S$_2$ alloys exhibit three characteristic peaks in the region from 350 to 420 cm^{-1}, assigned to the WS$_2$-like E', MoS$_2$-like E', and A'$_1$ modes (Figure 6c). The relative intensities and positions of these peaks vary depending on the Mo/W ratio in crystal (Chen et al. 2013). The Raman spectra of this crystal change with the compositions (Figure 6d-6f). From the peak energies of the PL spectra, the Mo/W ratio in specific crystals is determined (Figure 6g).

Revealing the heteroatomic distributions in two-dimensional crystals is particularly critical for material engineering at atomic level. By statistics of the homo- and heteroatomic coordinates in single-layered Mo$_{1-x}$W$_x$S$_2$ from the atomically resolved scanning transmission electron microscope images, the degree of alloying for the transition metal elements (Mo or W) can be successfully quantified. Such a direct route to gain an insight into the alloying degree on individual atom basis will find broad applications in characterizing low-dimensional heterocompounds and become an important complement to the existing theoretical methods (Dumcenco et al. 2013).

Figure 6. (a) Optical microscopy image and (b) structural model of the lateral and stacked heterostructure based on monolayer $Mo_{1-x}W_xS_2$ alloys grown at 750 °C. For the structural model in (b), cyan and red colors correspond to high populations of W and Mo atoms, respectively. (c) Raman spectra acquired at the points marked by solid circles in (f). (d) WS_2-like and (e) MoS_2-like E' Raman intensity maps, (f) combined Raman intensity maps from (d) and (e). (g) PL spectra acquired at the points marked by solid circles in (j). PL intensity maps from (h) 1.92 to 1.99 eV and (i) 1.80 to 1.88 eV, (j) combined PL intensity maps from (h) and (i). The order of the spectra (from top to bottom) corresponds to the positions of the circles in (f) and (j). The Raman and PL intensities are normalized to the maximum recorded intensities.
Source: Reprinted with permission from Kobayashi et al., (2015). Copyright of Tsinghua University Press and Springer-Verlag Berlin Heidelberg.

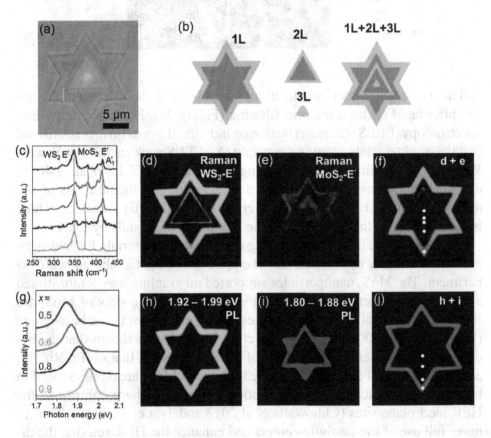

For example, the atomic arrangements of Mo and W atoms in $Mo_{1-x}W_xS_2$ monolayer alloys (Figure 7) were imaged by scanning transmission electron microscopy (STEM) (Chen et al. 2013). As shown in Figure 7b, due to its large Z number, the W atom has a larger annular dark-field (ADF) contrast

Figure 7. Structure of $Mo_{1-x}W_xS_2$ monolayer. (a) Top view of $Mo_{1-x}W_xS_2$ monolayer and a side view of a unit cell. (b) STEM image of $Mo_{0.47}W_{0.53}S_2$ monolayer. (c) STEM image in panel (b) after FFT filtering, showing two types of atoms with different contrasts. (d) EELS of two individual atoms indicated as "1" and "2" in panel (c), showing EELS characteristics of W (top) and Mo (bottom), respectively. The blue lines are original EELS spectra, and the green lines are spectra after background subtraction.
Source: Reprinted with permission from Chen et al., 2013. Copyright of American Chemical Society.

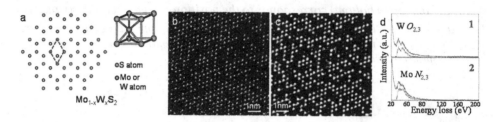

and then is brighter than the Mo atom, which is more clearly seen in the image after fast Fourier transform filtering (Figure 7c). Electron energy loss spectroscopy (EELS) characterization on individual atoms further confirmed that the brighter spots, showing two split $O_{2,3}$ EELS peaks at 40-50 eV, were W atoms and the dimer spots, showing one broad $N_{2,3}$ EELS peak at 40-50 eV, were Mo atoms (Figure 7d), By direct counting of the numbers of Mo and W atoms in STEM images, the W content x was directly calculated which is consistent with the W content value x of the bulk crystals by EDX.

Hinnemann et al. (2005) searching new catalytic materials for hydrogen evolution reaction using quantum chemical methods. Combined with experiment, The MoS_2 nanoparticles supported on graphite was demonstrated a new class of electrode materials (Figure 8). The edge sites of MoS_2 are found to have metallic electronic states, which are absent in the basal plane (Bollinger et al. 2001 and Lauritsen et al. 2007). It was shown that the hydrogen binding energy on the edge of MoS_2 is close to that on Pt, Rh, Re, and Ir (Jaramillo et al. 2007). These edge sites of MoS_2 are demonstrated to be active centers for hydrogen evolution reaction, in sharp contrast to the HER inert plane sites (Chhowalla et al 2013 and Tsai et al. 2014). Thus to make full use of the catalytic centers and enhance the HER activity, the dimension of the lateral direction of Mo(W) dicalcogenides layers should be significantly shrunk to increase the ratio of edge sites to planar sites (Bond et al. 2009).

It is generally accepted that HER reaction under acidic conditions consists of the three steps (Conway and Tilak 2002):

Synthesis, Characterization, and Catalytic Application of 2D Mo(W) Dichalcogenides

Figure 8. (a) Theoretical simulation of the adsorption free energies on different materials. (b) Hydrogen is bound at the Mo edge of MoS_2 slab. (c) Polarization curve for hydrogen evolution. (d) STM images of MoS_2 nanoparticles on modified graphite.
Source: Adapted with permission from Hinnemann et al., 2005. Copyright of American Chemical Society.

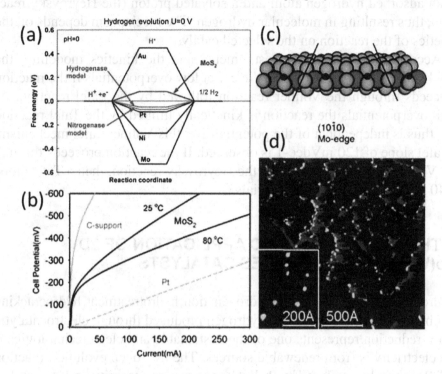

1. Proton adsorption with charge transfer (Volmer reaction):

$$H_3O^+ + e^- + S \rightarrow SH_{ads} + H_2O \tag{1}$$

2. Combination of surface hydrogen atoms (Tafel reaction):

$$SH_{ads} + SH_{ads} \rightarrow 2S + H_2 \tag{2}$$

3. Combination of a surface hydrogen atom and a solvated proton (Heyrovsky reaction):

$$SH_{ads} + H_3O^+ + e^- \rightarrow S + H_2 + H_2O \tag{3}$$

In Equations (1)-(3), S represents the free active sites on surface of electro-catalyst, while SHads denotes the adsorbed hydrogen atoms on the adsobed site. The HER begins with electrochemical adsorption of protons (the Volmer reaction), and then it is continued either with the combination of two adsorbed hydrogen atoms (the Tafel reaction) or with the combination of an adsorbed hydrogen atom and a solvated proton (the Heyrovsky reaction), thus resulting in molecular hydrogen. The mechanism depends on the kinetics of the reaction on the selected catalyst.

According to previous measurements and the kinetics modeling, the Tafel slope for the HER is 30 mVdec-1 at low overpotentials if the reaction proceeds through the Volmer reaction followed by the Tafel reaction. At high overpotentials the reaction is kinetically limited by the Tafel reaction and thus is independent of the potential. For this Volmer-Tafel mechanism, a Tafel slope of 120 mVdec-1 is observed. If the reaction proceeds through the Volmer reaction followed by the Heyrovsky reaction, then a Tafel slope is 40 mVdec-1 at low overpotentials.

5. THE NOVEL CATALYTIC APPLICATION OF 2D Mo(W) DICHALCOGENIDES CATALYSTS

Hydrogen is the basic feed in the hydrogenation, hydrotreatment, hydrocracking and hydrogenolysis processes. Hydrogen produced through electrocatalytic water reduction represents one of the sustainable and clean technologies if the electricity is from renewable sources. The hydrogen evolution reaction (HER) is usually catalyzed by Pt, but its expense and scarcity limit scalability opportunities. Thus noble metal free catalysts have attracted extensive interests. MoS_2 is evidenced a potential low-cost HER alternative to the traditional noble metal catalysts (Kibsgaard et al. 2012). Nanostructured MoS_2 is reported to be active and stable in acidic media (Hinneman et al. 2005). Amorphous MoS_x (Benck et al. 2012, Vrubel et al. 2012, Chang et al. 2013 and Li et al. 2014) and molecular mimics of the MoS_2 active site (Karunadasa et al. 2012) have also been explored as efficient HER catalysts. MoO_3-MoS_2 core-shell nanowires were demonstrated the impressive HER activity (Chen et al. 2013). Moreover, 1T-MoS_2 appears to be a more active hydrogen evolution catalyst than 2H-MoS_2 (Wang et al. 2013c). This is partly ascribed to the excellent electronic conduction of 1T-MoS_2. Monolayer nanosheets of chemically exfoliated WS_2 was reported an efficient catalyst for HER with very low overpotentials (Voiry et al. 2013a). Analyses indicated that the enhanced electrocatalytic activity of WS_2 is associated with the high concentration of the strained metallic 1T (octahedral) phase in the nanosheets.

Solar water splitting with semiconductor photo-catalysts is deemed promising also. It is well known that the photocatalytic H_2 evolution heavily relies on the separation rate of photogenerated electron-hole pairs and their followed interfacial catalytic reactions (Schneider et al. 2014). However, it is usually impossible to develop a high-efficiency photo-catalyst using a single semiconductor due to the rapid recombination of electrons and holes after light absorption. Electron-cocatalyst deposited on a photo-catalyst surface has been demonstrated to be an efficient strategy for the enhancement of photocatalytic performance via rapidly transferring interfacial electrons, retarding the recombination of photo-excited charges and providing effective active sites. Recently, many reports indicated that molybdenum sulfide could function as an electron-cocatalyst in photocatalytic hydrogen evolution reaction (Laursen et al 2012). Kanda et al. (2011) reported that the photocatalytic H_2 production of TiO_2 could be obviously improved by loading molybdenum sulfide nanoparticles. Xiang et al. (2012) observed that molybdenum sulfide modified rGO-TiO_2 composites exhibited an obviously higher hydrogen production activity than the rGO-TiO_2 or TiO_2. Yu et al. (2016) found that the H_2-evolution activity of g-C_3N_4 photocatalyst was obviously improved by loading amorphous MoS_x cocatalyst. Chang et al. (2014 and 2015) suggested that the unsaturated S atoms on the crystalline molybdenum sulfide edge rapidly capture protons from solution and promote the direct reduction of H_+ to H_2 by photogenerated electrons. They demonstrated that the photocatalytic performance of $nMoS_2$/CdS (where n rep-resents the layer numbers of MoS_2) had a significant increase with the gradually decreasing layer numbers of MoS_2 due to that the single or few layer MoS_2 have more unsaturated active S atoms.

Electrochemical or photochemical reduction of carbon dioxide (CO_2) could recycle the greenhouse gas back into fuels or chemicals. However, most catalysts are too inefficient in practice due to weak binding interactions between the reaction intermediates. And the catalyst gives rise to high overpotentials or slow electron transfer kinetics resulting in low exchange current densities. Asadi et al. (2014) recently reported that three-dimensional (3D) bulk molybdenum disulfide (MoS_2) catalyzes CO_2 reduction to CO at an extremely low overpotential (54 mV) in an ionic liquid (IL). They further demonstrated that 2D Mo and W dichalcogenides manifest even higher performance for electrocatalytic CO_2 reduction in the IL 1-ethyl-3-methylimidazolium tetrafluoroborate (EMIM-BF4). The tungsten diselenide nanosheets show a current density of 18.95 milliamperes per square centimeter, CO faradaic efficiency of 24%, and CO formation turnover frequency of 0.28 per second at a low overpotential of 54 millivolts. This exceptional performance

Figure 9. CO_2 reduction performance of the MoS_2 and WSe_2 catalysts, Ag NPs, and bulk Ag in the EMIM-BF4 solution. (A) Cyclic voltammetry (CV) curves for WSe_2 nano flakes (NFs), bulk MoS_2, Ag nanoparticles (Ag NPs), and bulk Ag in CO_2 environment. Inset shows the current densities in low overpotentials. (B) CO and H_2 overall faradaic efficiency (FE) at different applied potentials for WSe_2 NFs. The error bars represent SD of four measurements. (C) CO formation TOF of WSe2 NFs, bulk MoS_2, and Ag NPs in IL electrolyte at overpotentials of 54 to 650 mV. At 54 mV overpotential, Ag NPs' result is zero. (D) Overview of different catalysts' performance at different overpotentials (h).
Source: Reprinted with permission from Asadi et al. (2016) Copyright of American Association for the Advancement of Science.

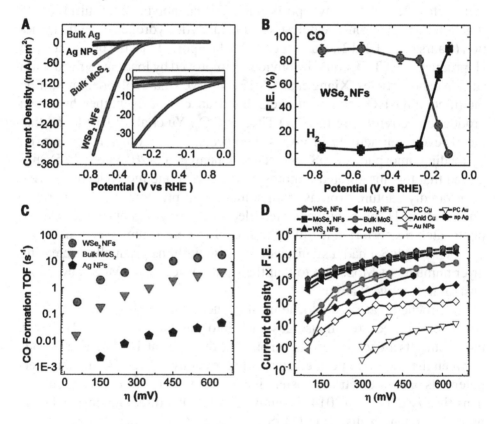

are attributed to a combination of low overpotentials and efficient electron transfer properties of the 2D nano sheets and the IL-enhanced local CO_2 concentration (Asadi et al. 2016).

Although the significant catalytic potential has been demonstrated for 2D layer Mo(W) dichalcogenides, there are considerable challenges in transferring this potential into practical technologies. One of challenge is the scalable production of these 2D materials with efficient and low-cost technique. Although we have seen increased efforts in developing new synthetic strategies for the large growth of these 2D materials recent years, however, the progress to date is still far from ideal.

Now, the catalytic explorations of 2D layer Mo(W) dichalcogenides are mainly focused on the HER and CO_2 reduction. However, there are huge chances to produce other fuels or chemicals with these 2D catalysts which needs a lot of efforts from fundamental researchers and application engineers.

REFERENCES

Afanasiev, P., Geantet, C., Llorens, I., & Proux, O. (2012). Biotemplated synthesis of highly divided MoS_2 catalysts. *Journal of Materials Chemistry*, *22*(19), 9731–9737. doi:10.1039/c2jm30377a

Asadi, M., Kim, K., Liu, C., Addepalli, A. V., Abbasi, P., Yasaei, P., & Sslehi-Khojin, A. et al. (2016). Nanostructured transition metal dichalcogenide electrocatalysts for CO_2 reduction in ionic liquid. *Science*, *353*(6298), 467–470. doi:10.1126/science.aaf4767 PMID:27471300

Asadi, M., Kumar, B., Behranginia, A., Rosen, B. A., Baskin, A., Repnin, N., & Salehi-Khojin, A. et al. (2014). Robust carbon dioxide reduction on molybdenum disulphide edges. *Nature Communications*, *5*, 4470–4478. doi:10.1038/ncomms5470 PMID:25073814

Benck, J. D., Chen, Z., Kuritzky, L. Y., Forman, A. J., & Jaramillo, T. F. (2012). Amorphous molybdenum sulfide catalysts for electrochemical hydrogen production: Insights into the origin of their catalytic activity. *ACS Catalysis*, *2*(9), 1916–1923. doi:10.1021/cs300451q

Bollinger, M. V., Lauritsen, J. V., Jacobsen, K. W., Nørskov, J. K., Helveg, S., & Besenbacher, F. (2001). One-dimensional metallic edge states in MoS_2. *Physical Review Letters*, *87*(19), 196803–196807. doi:10.1103/PhysRevLett.87.196803 PMID:11690441

Bonde, J., Moses, P. G., Jaramillo, T. F., Nørskov, J. K., & Chorkendorff, I. (2009). Hydrogen evolution on nano-particulate transition metal sulfides. *Faraday Discussions*, *140*, 219–231. doi:10.1039/B803857K PMID:19213319

Chang, K., Li, M., Wang, T., Ouyang, S., Li, P., Liu, L., & Ye, J. (2015). Drastic layer-number-dependent activity enhancement in photocatalytic H_2 evolution over $nMoS_2$/CdS (n ≥ 1) under visible light. *Advanced Energy Materials*, 5(10), 1402279. doi:10.1002/aenm.201402279

Chang, K., Mei, Z., Wang, T., Kang, Q., Ouyang, S., & Ye, J. (2014). MoS_2/Graphene cocatalyst for efficient photocatalytic H2 evolution under visible light irradiation. *ACS Nano*, 8(7), 7078–7087. doi:10.1021/nn5019945 PMID:24923678

Chang, Y. H., Lin, C. T., Chen, T. Y., Hsu, C. L., Lee, Y. H., Zhang, W., & Li, L. J. et al. (2013). Highly Efficient Electrocatalytic Hydrogen Production by MoS_x Grown on Graphene-Protected 3D Ni Foams. *Advanced Materials*, 25(5), 756–760. doi:10.1002/adma.201202920 PMID:23060076

Chen, X. J., Dobson, J. F., & Raston, C. L. (2012). Vortex fluidic exfoliation of graphite and boron nitride. *Chemical Communications*, 48(31), 3703–3705. doi:10.1039/c2cc17611d PMID:22314550

Chen, Y. F., Xi, J. Y., Dumcenco, D. O., Liu, Z., Suenaga, K., Wang, D., & Xie, L. M. et al. (2013). Tunable band gap photoluminescence from atomically thin transition-metal dichalcogenide alloys. *ACS Nano*, 7(5), 4610–4616. doi:10.1021/nn401420h PMID:23600688

Chen, Z., Cummins, D., Reinecke, B. N., Clark, E., Sunkara, M. K., & Jaramillo, T. F. (2011). Core-shell MoO_3-MoS_2 nanowires for hydrogen evolution: A functional design for electrocatalytic materials. *Nano Letters*, 11(10), 4168–4175. doi:10.1021/nl2020476 PMID:21894935

Chhowalla, M., Shin, H. S., Eda, G., Li, L. J., Loh, K. P., & Zhang, H. (2013). The chemistry of two-dimensional layered transition metal dichalcogenide nanosheets. *Nature Chemistry*, 5(4), 263–275. doi:10.1038/nchem.1589 PMID:23511414

Coleman, J. N., Lotya, M., ONeil, A., Bergin, S. D., King, P. J., Khan, U., & Nicolosi, V. et al. (2011). Two-dimensional nanosheets produced by liquid exfoliation of layered materials. *Science*, 331(6017), 568–571. doi:10.1126/science.1194975 PMID:21292974

Conway, B. E., & Tilak, B. V. (2002). Interfacial processes involving electrocatalytic evolution and oxidation of H_2, and the role of chemisorbed H. *Electrochimica Acta*, 47(22-23), 3571–3594. doi:10.1016/S0013-4686(02)00329-8

Cunningham, G., Lotya, M., Cucinotta, C. S., Sanvito, S., Bergin, S. D., Menzel, R., & Coleman, J. N. et al. (2012). Solvent exfoliation of transition metal dichalcogenides: Dispersibility of exfoliated nanosheets varies only weakly between compounds. *ACS Nano*, *6*(4), 3468–3480. doi:10.1021/nn300503e PMID:22394330

Duan, X., Wang, C., Pan, A., Yu, R., & Duan, X. (2015). Two-dimensional transition metal dichalcogenides as atomically thin semiconductors: Opportunities and challenges. *Chemical Society Reviews*, *44*(24), 8859–8876. doi:10.1039/C5CS00507H PMID:26479493

Duan, X., Wang, C., Shaw, J. C., Cheng, R., Chen, Y., Li, H., & Duan, X. et al. (2014). Lateral epitaxial growth of two-dimensional layered semiconductor heterojunctions. *Nature Technology*, *9*, 1024–1030. PMID:25262331

Dumcenco, D. O., Kobayashi, H., Liu, Z., Huang, Y.-S., & Suenaga, K. (2012). Visualization and quantification of transition metal atomic mixing in $Mo_{1-x}W_xS_2$ single layers. *Nature Communications*, *4*, 1351–1356. doi:10.1038/ncomms2351 PMID:23322039

Dunne, P. W., Munn, A. S., Starkey, C. L., & Lester, E. H. (2015). The sequential continuous-flow hydrothermal synthesis of molybdenum disulphide. *Chemical Communications*, *51*(19), 4048–4050. doi:10.1039/C4CC10158H PMID:25660109

Fang, H., Battaglia, C., Carraro, C., Nemsak, S., Ozdol, B., & Kang, J. S., ... Javey, A. (2014). Strong interlayer coupling in van der Waals heterostructures built from single-layer chalcogenides. *Proceedings of the National Academy of Sciences of the United States of America*, *111*, 6198–6202. doi:10.1073/pnas.1405435111

Frindt, R. F. (1966). Single crystals of MoS_2 several molecular layers thick. *Journal of Applied Physics*, *37*(4), 1928–1929. doi:10.1063/1.1708627

Geim, A. K., & Grigorieva, I. V. (2013). Van der Waals heterostructures. *Nature*, *499*(7459), 419–425. doi:10.1038/nature12385 PMID:23887427

Halim, U., Zheng, C. R., Chen, Y., Lin, Z., Jiang, S., Cheng, R., & Duan, X. et al. (2013). A rational design of cosolvent exfoliation of layered materials by directly probing liquid–solid interaction. *Nature Communications*, *4*, 2213. doi:10.1038/ncomms3213 PMID:23896793

Heine, T. (2015). Transition metal chalcogenides: Ultrathin inorganic materials with tunable electronic properties. *Accounts of Chemical Research*, *48*(1), 65–72. doi:10.1021/ar500277z PMID:25489917

Heising, J., & Kanatzidis, M. G. (1999). Structure of restacked MoS_2 and WS_2 elucidated by electron crystallography. *Journal of the American Chemical Society*, *121*(4), 638–643. doi:10.1021/ja983043c

Hinnemann, B., Moses, P. G., Bonde, J., Jørgensen, K. P., Nielsen, J. H., Horch, S., & Nørskov, J. K. et al. (2005). Biomimetic Hydrogen Evolution: MoS_2 Nanoparticles as Catalyst for Hydrogen Evolution. *Journal of the American Chemical Society*, *127*(15), 5308–5309. doi:10.1021/ja0504690 PMID:15826154

Huang, X., Zeng, Z., Bao, S., Wang, M., Qi, X., Fan, Z., & Zhang, H. (2013). Solution-phase epitaxial growth of noble metal nanostructures on dispersible single-layer molybdenum disulfide nanosheets. *Nature Communications*, *4*, 1444. doi:10.1038/ncomms2472 PMID:23385568

Jaramillo, T. F., Jørgensen, K. P., Bonde, J., Nielsen, J. H., Horch, S., & Chorkendorff, I. (2007). Identification of active edge sites for electrochemical H_2 evolution from MoS_2 nanocatalysts. *Science*, *317*(5834), 100–102. doi:10.1126/science.1141483 PMID:17615351

Jeffery, A. A., Nethravathi, C., & Rajamathi, M. (2014). Two-dimensional nanosheets and layered hybrids of MoS_2 and WS_2 through exfoliation of ammoniated MS_2 (M = Mo,W). *The Journal of Physical Chemistry C*, *118*(2), 1386–1396. doi:10.1021/jp410918c

Joensen, P., Frindt, R. F., & Morrison, S. R. (1986). Single-layer MoS_2. *Materials Research Bulletin*, *21*(4), 457–461. doi:10.1016/0025-5408(86)90011-5

Kanda, S., Akita, T., Fujishima, M., & Tada, H. (2011). Facile synthesis and catalytic activity of MoS_2/TiO_2 by a photodeposition-based technique and its oxidized derivative MoO_3/TiO_2 with a unique photochromism. *Journal of Colloid and Interface Science*, *354*(2), 607–610. doi:10.1016/j.jcis.2010.11.007 PMID:21167494

Kang, K., Xie, S., Huang, L., Han, Y., Huang, P. Y., Mak, K. F., & Park, J. et al. (2015). High-mobility three-atom-thick semiconducting films with wafer-scale homogeneity. *Nature*, *520*(7549), 656–660. doi:10.1038/nature14417 PMID:25925478

Karunadasa, H. I., Montalvo, E., Sun, Y., Majda, M., Long, J. R., & Chang, C. J. (2012). A Molecular MoS_2 Edge Site Mimic for Catalytic Hydrogen Generation. *Science, 335*(6069), 698–702. doi:10.1126/science.1215868 PMID:22323816

Kibsgaard, J., Chen, Z., Reinecke, B. N., & Jaramillo, T. F. (2012). Engineering the surface structure of MoS_2 to preferentially expose active edge sites for electrocatalysis. *Nature Materials, 11*(11), 963–969. doi:10.1038/nmat3439 PMID:23042413

Kobayashi, Y., Mori, S., Maniwa, Y., & Miyata, Y. (2015). Bandgap-tunable lateral and vertical heterostructures based on monolayer $Mo_{1-x}W_xS_2$ alloys. *Nano Research, 8*(10), 3261–3271. doi:10.1007/s12274-015-0826-7

Komsa, H.-P., & Krasheninnikov, A. V. (2012). Two-Dimensional Transition Metal Dichalcogenide Alloys: Stability and Electronic Properties. *The Journal of Physical Chemistry Letters, 3*(23), 3652–3656. doi:10.1021/jz301673x PMID:26291001

Kong, D., Wang, H., Cha, J. J., Pasta, M., Koski, K. J., Yao, J., & Cui, Y. (2013). Synthesis of MoS_2 and $MoSe_2$ Films with Vertically Aligned Layers. *Nano Letters, 13*(3), 1341–1347. doi:10.1021/nl400258t PMID:23387444

Lauritsen, J. V., Kibsgaard, J., Helveg, S., Topsoe, H., Clausen, B. S., Laegsgaard, E., & Besenbacher, F. (2007). Size-dependent structure of MoS_2 nanocrystals. *Nature Nanotechnology, 2*(1), 53–58. doi:10.1038/nnano.2006.171 PMID:18654208

Laursen, A. B., Kegnaes, S., Dahl, S., & Chorkendorff, I. (2012). Molybdenum sulfides-efficient and viable materials for electro - and photoelectrocatalytic hydrogen evolution. *Energy & Environmental Science, 5*(2), 5577–5591. doi:10.1039/c2ee02618j

Lee, Y. H., Yu, L. L., Wang, H., Fang, W. J., Ling, X., Shi, Y. M., & Kong, J. et al. (2013). Synthesis and transfer of single-layer transition metal disulfides on diverse surfaces. *Nano Letters, 13*(4), 1852–1857. doi:10.1021/nl400687n PMID:23506011

Li, D. J., Maiti, U. N., Lim, J., Choi, D. S., Lee, W. J., Oh, Y., & Kim, S. O. et al. (2014b). Molybdenum Sulfide/N-Doped CNT Forest Hybrid Catalysts for High-Performance Hydrogen Evolution Reaction. *Nano Letters, 14*(3), 1228–1233. doi:10.1021/nl404108a PMID:24502837

Li, H., Wu, J., Yin, Z., & Zhang, H. (2014a). Preparation and Applications of Mechanically Exfoliated Single-Layer and Multilayer MoS_2 and WSe_2 Nanosheets. *Accounts of Chemical Research, 47*(4), 1067–1075. doi:10.1021/ar4002312 PMID:24697842

Li, H., Zhang, Q., Duan, X., Wu, X., Fan, X., Zhu, X., & Pan, A. et al. (2015). Lateral Growth of Composition Graded Atomic Layer $MoS_{2(1-x)}Se_{2x}$ Nanosheet. *Journal of the American Chemical Society, 137*(16), 5284–5287. doi:10.1021/jacs.5b01594 PMID:25871953

Li, Y., Wang, H., Xie, L., Liang, Y., Hong, G., & Dai, H. (2011). MoS_2 Nanoparticles Grown on Graphene: An Advanced Catalyst for the Hydrogen Evolution Reaction. *Journal of the American Chemical Society, 133*(19), 7296–7299. doi:10.1021/ja201269b PMID:21510646

Lin, Y.-C., Dumcencon, D. O., Huang, Y.-S., & Suenaga, K. (2014). Atomic Mechanism of the semiconducting -to-Metallic Phase Transition in Single-Layered MoS_2. *Nature Nanotechnology, 9*(5), 391–396. doi:10.1038/nnano.2014.64 PMID:24747841

Lin, Y. C., Lu, N., Lopez, N. P., Li, J., Lin, Z., Peng, X., & Robinson, J. A. et al. (2014). Direct synthesis of van der Waals solids. *ACS Nano, 8*(4), 3715–3723. doi:10.1021/nn5003858 PMID:24641706

Liu, K. K., Zhang, W. J., Lee, Y. H., Lin, Y. C., Chang, M. T., Su, C., & Li, L. J. et al. (2012). Growth of Large-Area and Highly Crystalline MoS_2 Thin Layers on Insulating Substrates. *Nano Letters, 12*(3), 1538–1544. doi:10.1021/nl2043612 PMID:22369470

Lu, Z., Wang, H., Kong, D., Yan, K., Hsu, P.-C., Zheng, G., & Cui, Y. et al. (2014). Electrochemical tuning of layered lithium transition metal oxides for improvement of oxygen evolution reaction. *Nature Communications, 5*, 4345. PMID:24993836

Lukowski, M. A., Daniel, A. S., English, C. R., Meng, F., Forticaux, A., Hamers, R. J., & Jin, S. (2014). Highly active hydrogen evolution catalysis from metallic WS_2 nanosheets. *Energy & Environmental Science, 7*(8), 2608–2613. doi:10.1039/C4EE01329H

Lukowski, M. A., Daniel, A. S., Meng, F., Forticaux, A., Li, L., & Jin, S. (2013). Enhanced hydrogen evolution catalysis from chemically exfoliated metallic MoS_2 nanosheets. *Journal of the American Chemical Society, 135*(28), 10274–10277. doi:10.1021/ja404523s PMID:23790049

Lv, R., Robinson, J. A., Schaak, R. E., Sun, D., Sun, Y., Mallouk, T. E., & Terrones, M. (2015). Transition Metal Dichalcogenides and Beyond: Synthesis, Properties, and Applications of Single- and Few-Layer Nanosheets. *Accounts of Chemical Research*, *48*(1), 56–64. doi:10.1021/ar5002846 PMID:25490673

Mak, K. F., He, K., Shan, J., & Heinz, T. F. (2012). Control of valley polarization in monolayer MoS_2 by optical helicity. *Nature Nanotechnology*, *7*(8), 494–498. doi:10.1038/nnano.2012.96 PMID:22706698

Mann, J., Ma, Q., Odenthal, P. M., Isarraraz, M., Le, D., Preciado, E., & Bartels, L. et al. (2014). 2-Dimensional Transition Metal Dichalcogenides with Tunable Direct Band Gaps: $MoS_{2(1-x)}Se_{2x}$ Monolayers. *Advanced Materials*, *26*(9), 1399–1404. doi:10.1002/adma.201304389 PMID:24339159

Najmaei, S., Liu, Z., Zhou, W., Zou, X., Shi, G., Lie, S., & Lou, J. et al. (2013). Vapour phase growth and grain boundary structure of molybdenum disulphide atomic layers. *Nature Materials*, *12*(8), 754–759. doi:10.1038/nmat3673 PMID:23749265

Nguyen, E. P., Carey, B. J., Daeneke, T., Ou, J. Z., Latham, K., Zhuiykov, S., & Kalantar-zadeh, K. (2015). Investigation of two-Solvent grinding-assisted liquid phase exfoliation of layered MoS_2. *Chemistry of Materials*, *27*(1), 53–59. doi:10.1021/cm502915f

Nicolosi, V., Chhowalla, M., Kanatzidis, M. G., Strano, M. S., & Coleman, J. N. (2013). Liquid exfoliation of layered materials. *Science*, *340*(6139), 1421–1439. doi:10.1126/science.1226419 PMID:23661643

Novoselov, K. S., Geim, A. K., Morozov, S. V., Jiang, D., Zhang, Y., Dubonos, S. V., & Firsov, A. A. et al. (2004). Electric field effect in atomically thin carbon films. *Science*, *306*(5696), 666–669. doi:10.1126/science.1102896 PMID:15499015

Osada, M., & Sasaki, T. (2012). Two-dimensional dielectric nanosheets: Novel nanoelectronics from nanocrystal building blocks. *Advanced Materials*, *24*(2), 210–228. doi:10.1002/adma.201103241 PMID:21997712

Ramakrishna Matte, H. S. S., Gomathi, A., Manna, A. K., Late, D. J., Datta, R., Pati, S. K., & Rao, C. N. R. (2010). MoS2 and WS2 Analogues of Graphene. *Angewandte Chemie*, *122*(24), 4153–4156. doi:10.1002/ange.201000009

Sandoval, S. J., Yang, D., Frindt, R. F., & Irwin, J. C. (1991). Raman-Study and Lattice-Dynamics of Single Molecular Layers of MoS_2. *Physical Review B: Condensed Matter and Materials Physics, 44*(8), 3955–3962. doi:10.1103/PhysRevB.44.3955 PMID:10000027

Schneider, J., Matsuoka, M., Takeuchi, M., Zhang, J., Horiuchi, Y., Anpo, M., & Bahnemann, D. W. (2014). Understanding TiO2 Photocatalysis: Mechanisms and Materials. *Chemical Reviews, 114*(19), 9919–9986. doi:10.1021/cr5001892 PMID:25234429

Shaw, J. C., Zhou, H., Chen, Y., Weiss, N. O., Liu, Y., Huang, Y., & Duan, X. (2014). Chemical vapor deposition growth of monolayer $MoSe_2$ nanosheets. *Nano Research, 7*(4), 511–517. doi:10.1007/s12274-014-0417-z

Terrones, H., López-Urías, F., & Terrones, M. (2013). Novel hetero-layered materials with tunable direct band gaps by sandwiching different metal disulfides and diselenides. *Scientific Reports, 3*, 1549. doi:10.1038/srep01549 PMID:23528957

Tsai, C., Chan, K., Abild-Pedersen, F., & Nørskov, J. K. (2014). Active edge sites in MoSe2 and WSe2 catalysts for the hydrogen evolution reaction: A density functional study. *Physical Chemistry Chemical Physics, 16*(26).

van der Zande, A. M., Huang, P. Y., Chenet, D. A., Berkelbach, T. C., You, Y., Lee, G., & Hone, J. C. et al. (2013). Grains and grain boundaries in highly crystalline monolayer molybdenum disulphide. *Nature Materials, 12*(6), 554–561. doi:10.1038/nmat3633 PMID:23644523

Varrla, E., Backes, C., Paton, K. R., Harvey, A., Gholamvand, Z., McCauley, J., & Coleman, J. N. (2015). Large-scale production of size-controlled MoS2 nanosheets by shear exfoliation. *Chemistry of Materials, 27*, 1129–1139.

Voiry, D., Salehi, M., Silva, R., Fujita, T., Chen, M., Asefa, T., & Chhowalla, M. et al. (2013a). Conducting MoS_2 Nanosheets as Catalysts for Hydrogen Evolution Reaction. *Nano Letters, 13*(12), 6222–6227. doi:10.1021/nl403661s PMID:24251828

Voiry, D., Yamaguchi, H., Li, J., Silva, R., Alves, D. C. B., Fujita, T., & Chhowalla, M. et al. (2013b). Enhanced catalytic activity in strained chemically exfoliated WS_2 nanosheets for hydrogen evolution. *Nature Materials, 12*(9), 850–855. doi:10.1038/nmat3700 PMID:23832127

Vrubel, H., Merki, D., & Hu, X. (2012). Hydrogen evolution catalyzed by MoS_3 and MoS_2 particles. *Energy & Environmental Science*, *5*(3), 6136. doi:10.1039/c2ee02835b

Wang, H., Kong, D., Johanes, P., Cha, J. J., Zheng, G., Yan, K., & Cui, Y. et al. (2013a). $MoSe_2$ and WSe_2 Nanofilms with Vertically Aligned Molecular Layers on Curved and Rough Surfaces. *Nano Letters*, *13*(7), 3426–3433. doi:10.1021/nl401944f PMID:23799638

Wang, H., Lu, Z., Kong, D., Sun, J., Hymel, T. M., & Cui, Y. (2014). Electrochemical Tuning of MoS_2 Nanoparticles on Three-Dimensional Substrate for Efficient Hydrogen Evolution. *ACS Nano*, *8*(5), 4940–4947. doi:10.1021/nn500959v PMID:24716529

Wang, H., Lu, Z., Xu, S., Kong, K., Cha, J. J., & Zheng, G., ... Cui, Y. (2013b). Electrochemical tuning of vertically aligned MoS2 nanofilms and its application in improving hydrogen evolution reaction. *Proceedings of the National Academy of Sciences of the United States of America*, *110*, 19701–19706. doi:10.1073/pnas.1316792110

Wang, H., Yuan, H., Hong, S. S., Li, Y., & Cui, Y. (2015). Physical and chemical tuning of two-dimensional transition metal dichalcogenides. *Chemical Society Reviews*, *44*(9), 2664–2680. doi:10.1039/C4CS00287C PMID:25474482

Xiang, Q., Yu, J., & Jaroniec, M. (2012). Synergetic Effect of MoS_2 and Graphene as Cocatalysts for Enhanced Photocatalytic H_2 Production Activity of TiO_2 Nanoparticles. *Journal of the American Chemical Society*, *134*(15), 6575–6578. doi:10.1021/ja302846n PMID:22458309

Xiao, D., Liu, G.-B., Feng, W., Xu, X., & Yao, W. (2012). Coupled Spin and Valley Physics in Monolayers of MoS2 and Other Group-VI Dichalcogenides. *Physical Review Letters*, *108*(19), 196802. doi:10.1103/PhysRevLett.108.196802 PMID:23003071

Xie, J., Zhang, J., Li, S., Grote, F., Zhang, X., Zhang, H., & Xie, Y. et al. (2013). Controllable Disorder Engineering in Oxygen-Incorporated MoS_2 Ultrathin Nanosheets for Efficient Hydrogen Evolution. *Journal of the American Chemical Society*, *135*(47), 17881–17888. doi:10.1021/ja408329q PMID:24191645

Yao, Y., Tolentino, L., Yang, Z., Song, X., Zhang, W., Chen, Y., & Wong, C. (2013). High-Concentration Aqueous Dispersions of MoS_2. *Advanced Functional Materials*, *23*(28), 3577–3583. doi:10.1002/adfm.201201843

Yoosuk, B., Tumnantong, D., & Prasassarakich, P. (2012). Amorphous unsupported Ni–Mo sulfide prepared by one step hydrothermal method for phenol hydrodeoxygenation. *Fuel, 91*(1), 246–252. doi:10.1016/j.fuel.2011.08.001

Yu, H., Xiao, P., Wang, P., & Yu, J. (2016). Amorphous molybdenum sulfide as highly efficient electron-cocatalyst for enhanced photocatalytic H_2 evolution. *Applied Catalysis B: Environmental, 193*, 217–225. doi:10.1016/j.apcatb.2016.04.028

Zeng, Z., Sun, T., Zhu, J., Huang, X., Yin, Z., Lu, G., ... Zhang, H. (2012) An effective method for the fabrication of few-Layer-thick inorganic nanosheets. Angewandte Chemie-International Edition, 51, 9052–9056.

Zeng, Z., Tan, C., Huang, X., Bao, S., & Zhang, H. (2014). Growth of noble metal nanoparticles on single-layer TiS_2 and TaS_2 nanosheets for hydrogen evolution reaction. *Energy Environ. Sci., 7*(2), 797–803. doi:10.1039/C3EE42620C

Zeng, Z., Yin, Z., Huang, X., Li, H., He, Q., Lu, G., ... Zhang, H. (2011) Single-layer semiconducting nanosheets: High-yield preparation and device fabrication. Angewandte Chemie-International Edition, 50, 11093–11097.

Zhou, H., Wang, C., Shaw, J. C., Cheng, R., Chen, Y., Huang, X., & Duan, X. et al. (2015). Large Area Growth and Electrical Properties of p-Type WSe_2 Atomic Layers. *Nano Letters, 15*(1), 709–713. doi:10.1021/nl504256y PMID:25434747

Zhou, K.-G., Mao, N.-N., Wang, H.-X., Peng, Y., & Zhang, H.-L. (2011). A Mixed-Solvent Strategy for Efficient Exfoliation of Inorganic Graphene Analogues. *Angewandte Chemie International Edition, 50*(46), 10839–10842. doi:10.1002/anie.201105364 PMID:21954121

Chapter 2
The Fundamental Research and Application Progress of 2D Layer Mo(W)S$_2$-Based Catalyst

The Mo(W)S$_2$ based catalysts have been intensively studied by the global scientific community because of their vast implications in oil refining and chemical industry. The last two decades have seen major advances in the detailed understanding about their active sites at the atomic scale, the catalytic reaction mechanism, the interactions between active nanoparticles and supports, and so on. This newly acquired insight has opened up the way to a continual enhancement of the performance of Mo(W)S$_2$ based catalysts.

Mo(W)S$_2$ based catalysts are extensively used in the refining process for removal of impurities and enhancement of product quality. The global crude oils are becoming increasingly heavy and rich in heteroatoms (S, N, O, V, Ni, etc.). Meanwhile production of "clean fuels" has been identified a priority to meet more stringent legislation on the limitation of sulfur, olefin and aromatics contents. Facing with this challenge, a lot of efforts have been dedicated to constructing the connection between fundamental principles

DOI: 10.4018/978-1-5225-2274-4.ch002

and practical application of Mo(W)S_2 catalysts. And catalysts with better activity, selectivity and durability have been designed and developed based on the rich understanding of the synthesizing technique, active structure control and support modulation. And chemical application of Mo(W)S_2 based catalysts have attracted renewed interests owing to the fast advance in the micro-engineering technique in 2D materials which make it possible to finely control the catalytic selectivity. In this chapter, we review the recent advances in the fundamental research and practical application of the 2D layer Mo(W)S_2 based catalyst, and then project their potentials in improvement of the quality of fuels and production of new chemicals.

1. SUPPORTED 2D LAYER Mo(W)S_2 CATALYSTS

Supporting the Mo(W)S_2 nanosheeets on a high surface area material is beneficial to increases the number of active sites via enhanced dispersion and distribution of active phases. Meanwhile cheaper support decreases the cost of industry catalysts.

In petroleum and chemical industry, conventionally, the Mo(W)S_2 nanosheets is supported on the alumina support and promoted by Co or Ni co-catalysts. In the sulfiding process before usage, the Mo(W)S_2 active phases are yielded in situ by the sulfidation of the oxidic precursors, which are in advance prepared through three steps, i.e., impregnation, drying and calcination. The almost exclusive use of alumina as support is due to its outstanding textural and mechanical properties and the relatively low cost. However, the undesirable very strong metal-support interaction has urged to introduce new supports.

Recently, ordered mesoporous materials with high surface area, large pore volume, ordered pore structure, and good thermal and mechanical stabilities have been studies as the support of Mo(W)S_2 based nanosheets. They act as host to support the active species and/or behave as a nanoreactor to provide a space for the catalytic reactions. According to the pore symmetry, these meso materials can be classified as 2D (e.g. HMS, SBA-15 and MCM-41) and 3D (e.g. MCM-48, SBA-16, KIT-6, and FDU-12) architectures (Fan et al. 1998). When incorporated into the pores of the support, the active phases will be confined by the pore size and pore structure, evoking interesting modulation of the structural and electronic properties.

When MoS_2 nanosheets were supported on the FDU-12 mesoporous material, with the increasing of Mo loading, the layer structure of MoS_2 active phase transforms from straight to slightly curved then to ring-like and finally to spherical like morphology in the pores due to the restriction of the cage-like pores of FDU-12. And catalyst with 10 wt% MoO_3 loading shown the hydrodesulfurizaiton (HDS) performance of dibenzothiophens (DBT) superior to those supported on commercial γ-Al_2O_3 and SBA-15 (Liu et al 2016a).

SBA-16-supported ternary CoMoW catalysts were demonstrated to be more active in HDS of DBT reaction than their SBA-15-supported counterparts (Huirache-Acuña et al. 2009). It was attributed to that SBA-16 substrate possesses super large cage-like mesoporous structure with a high surface area and high thermal stability, thus providing favorable mass transfer kinetics.

For supported Co(Ni)MoW catalysts, Huirache-Acuña et al. (2012) compared the effects of the support (HMS versus SBA-16), promoter (Co vs. Ni) and modification of support with Al on the HDS of DBT. The results revealed that NiMoW/Al-HMS catalyst was more active than all Co-promoted catalysts (including a commercial CoMo/Al_2O_3 catalyst). The HDS activity was markedly influenced by the textural properties of support and the dispersion of the active phases. After Al doping, the presence of extra framework $AlNO_3$ phases on the surface of CoMoW/Al-SBA16 suppressed its HDS activity by formation of a less active "onion-type" $Mo(W)S_2$ structure.

Mendoza-Nieto et al. (2012) compared the HDS activities of NiMo catalysts supported on different types of silica supports (nanostructured supports of MCM-41 and SBA-15 types and commercial amorphous silica) and the same ones modified by TiO_2 grafting. Titania addition to all silica supports resulted in an increase in the HDS activity. However, this increase was smaller for the MCM-41 support than for the SBA-15 and amorphous silica supports. The NiMo/Ti-SBA-15 catalyst was found significantly more active (~40%) than the reference NiMo/γ-Al_2O_3 catalyst for HDS of 4,6-DMDBT. It revealed that the increased metal-support interaction after titania grafting improved the dispersion of Mo oxide species on Ti-SBA-15 support.

Recently, Liu et al. (2016b) synthesized a series of ordered mesoporous NiMo-Al_2O_3 catalysts via a solvent evaporation-induced self-assembly (EISA) method by using P123 as surfactant and anhydrous ethanol as solvent. Calcination of the as-synthesized NiMo-Al_2O_3 precursors at 500∘C can completely remove the surfactant and solvent, leading to the formation of hexagonally ordered mesoporous catalysts with a space group of p6mm. Characterizations evidenced that incorporations of active components (NiO and MoO_3) in the synthesis procedure maintained the ordered mesoporous structure and the active components were homogeneously distributed into the channels of alumina. After sulfidation, the ordered mesoporous NiMo-Al_2O_3 catalysts with the Ni/Mo molar ratio of 1:1 showed higher HDS activities of DBT than the impregnated NiMo/γ-alumina catalyst, and direct desulfurization (DDS) was the predominant pathway.

Laurenti et al. (2013) investigated the effect of three kinds of alumina supports (δ, γc and γT) on HDS. They observed a gain in HDS activity for unpromoted MoS_2 catalysts and CoMoS catalyst by using δ-alumina as a support instead of γ-alumina. This support effect is due to the better dispersion of the active phase on δ-alumina coupled with lower interactions between MoS_2 nanocrystallites and the support surface.

An efficient method to improvement of the properties of support is via modifying them with additives, such as boron, fluorine or phosphorous, during the preparation or in a post process. Such additives are used in order to control the surface acidities, the interaction between the support and the active phase, as well as the dispersion and structures of active components.

NiMo/Al_2O_3 catalyst doped with F showed an apparent improvement for vacuum residue hydroconversion. This is mostly due to the increase of the acidic and hydrogenating functions by F addition. And less coke was remained in spent catalyst, suggesting that coke precursors were cracked and hydrogenated (Marchal et al. 2012).

On the other hand, the conclusions on the HDS performance with P and B doping the alumina support are some contradictory in references (Guzmán et al. 2013). This is probably due to that the final effect of phosphorous or boron is the result of various cumulative effects. It was reported that phosphorous or boron incorporation can lead to: (1) a decrease of metal-support interaction; (2) a decrease of specific surface area and pore volume due to a pore plugging phenomenon; and (3) an increase of acidity. Therefore, the amount and type of additives should be optimized for each catalyst system in order to achieve needed results. Interestingly, the positive results for HDN are usually observed for P or B addition mainly due to the acidic sites facilitating the adsorption of N compounds (Vatutina et al. 2016).

2. THE DOPING METALS TO 2D Mo(W)S_2-BASED CATALYSTS

Co and Ni are the main promoters of Mo(W)S_2 based catalysts. Fe have also been used to a less extent. There are two models argued now to explain the promoted effects of transient metals on the HDS and hydrogenation activities. Edge decoration model attributed the high reaction activity of Ni or Co promoted MoS_2 to the formation of Co(Ni)-Mo-S phase (Topsøe et al. 1981 and 2007). This model have been extensively accepted to explain the promoting effects of bi-components sulfide catalysts and have been evidenced by STM technique (Lauritsen et al. 2007).

The Remote Control (RC) model (Li et al. 1997 and Delmaon and Froment 1996) proposed that hydrogen created on a donor phase (e.g. NiS_x or CoS_x) migrates to the acceptor (e.g. MoS_2 or WS_2) via spillover and enhance the hydrogenation abilities. The synergism between these two sulfides leads the high HDS catalytic activity of promoted sulfide catalysts (Villarroel et al. 2008 and Pimerzin et al. 2015). This model have been evidenced in some investigations where the formation of Co-Mo-S phase was physical prohibited using stacked bed systems (Ojeda et al. 2003). This model is mainly based on the investigation of unsupported sulfide catalyst. Wang et al. (2016) observed improved HDS performance of high promoter loading unsupported NiMo catalyst, which was ascribed to the spillover hydrogen effect between the NiS_x and MoS_2 two phases. Ramos et al. (2012) reported that the activity of bi-component sulfide catalyst was directly connected with an increase of the contact surface area between MoS_2 and Co_9S_8 phases.

It has long been debated between the Co(Ni)-Mo-S model and RC one. In fact, these two models all have their significances. Which one plays more important role depends on the preparation method and components of catalyst. And competition or synergy may exist in real HDS catalyst system.

In the early period, Harris (1986) found that the introduction of Cu in unsupported MoS_2 via reaction of $CuCl_2$ with $(NH_4)_2MoS_4$ only resulted a poison effect, because Cu decreased the electron density on Mo, thus leading to oxidization of Mo in MoS_2. Kibsgaard et al. (2010) revealed that Cu addition can form mixed-metal Cu-Mo-S type structures shaped as single-layer hexagonally truncated triangular MoS_2-like nanoclusters. Recently by preparation with chemical precipitation, the NiCuMo catalyst was shown improvement of HDS activity of 4,6-DMDBT and FCC diesel, The HRTEM photographs and XRD characterization showed the formation of smaller and more layers of MoS_2 nanoclusters. And TPR revealed that the reduction of nickel was shifted to lower temperature region resulting in the NiS_x nickel species after Cu introduction. A balance between Cu and Ni loadings is necessary to obtain the improved effect of HDS (Liu et al. 2014). It was observed in another investigation that the introduction of Zn enhanced the sulfidation of Mo species. The NiZnMo catalyst with the ratio of 9.5:0.5:10 showed a higher reaction rate of DBT than the other ratios of NiZnMo or Zn free catalysts. (Liu et al. 2015). Alkali doping to MoS_2 induce the hydrogenation of CO selectively to higher carbon alcohols. Increase of K loading enhances Mo dispersion and the number of active sites on MoS_2 domains (Surisetty et al., 2009 and 2011).

3. THE ADDITION OF CHELATING TO 2D Mo(W)S$_2$ -BASED CATALYSTS

Co(Ni)Mo(W)S$_2$ nanosheets supported on alumina lead to the formation of strong Mo-O-Al bonds, generating of type I CoMoS active phases with a low HDS activity (Topsøe, 2007). For avoiding this issue, a chelating agent assisted impregnation method without the calcination step was thereupon developed. Selective formation of the Co-Mo-S active phase was achieved by using a cobalt organometallic complexe, such as Co(acac)$_2$, Co(Cp)$_2$, or Co(CO)$_3$NO as a precursor (Laurenti et al. 2013 and Okamota et al. 2009). Chelating agents added with proper contents during the preparation of Co(Ni)Mo(W)S$_2$ catalysts show improved HDS activity (Sun et al. 2003 and Lélias et al. 2009). Other chelating agents such as ethylene diamine (EN), citric acid (CA), glycol, 1,2-cyclohexanediaminetetraacetic acid, nitrilotriacetic acid, and ethylenediaminetetraacetic acid have also been investigated (Rinaldi et al. 2009 and Mozhaev et al. 2016). They are considered to play three potential roles: 1) delay the sulfurization of Co or Ni species later than Mo or W species, to form more Co(Ni)-Mo(W)-S phases; 2) act as the ligands of Mo anions to form Mo complex, thus weakening the metal-support interaction and generating more type II active phases; 3) formation of Co-Mo-C active structures (Figure 1. Ge et al. 2015).

Rinaldi et al. (2010) and Mazoyer et al. (2008) developed a post-treatment method where the chelating agents are added into the calcined oxidic catalysts, the resulted catalysts hold more suitable metal-support interaction and

Figure 1. Schematic morphology and structure of CoMo catalyst prepared with addition of EN. Thiophene HDS activities over sulfided catalysts. Reaction conditions: t = 240 °C, p = 1.0 MPa, V(H$_2$)/V(oil) ≈ 200. Thiophene and DMDS sulfidations are designated -T and -D respectively.
Source: Adapted with permission from Ge et al. (2015). Copyright of American Chemical Society.

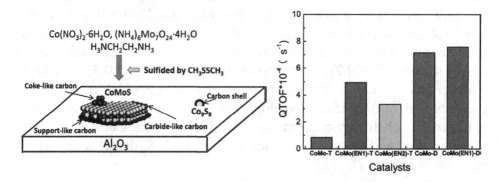

higher dispersion for metal species which promote formation of more type II Co-Mo-S active sites, thus enhancing the HDS activity (Zhang, 2016).

In order to answer the challenging environmental and energetic issues, developing highly active and selective $Mo(W)S_2$ based catalysts is of critical important. Improving preparation techniques using novel support materials and assisted agents can facilitate the development of industrial catalysts.

4. CHARACTERIZATION AND REACTION MECHANISM OF 2D $Mo(W)S_2$ BASED CATALYSTS

The layered structure of $Mo(W)S_2$ active phases consists of strong covalent bonding between the Mo and S atoms and the weak Van der Waals attraction between the lattice layers. When they are supported on alumina, due to the strong interactions of active phase with support, $Mo(W)S_2$ are highly dispersed with small size and single or several layers. However, their morphology can change in the real reaction atmosphere and temperature. The surface energy thermodynamically dominates this transform of morphologies. By the Wulff construction principal, the crystal will form morphology with the least surface energy (Walton et al. 2013).

The very powerful combination of STM experiment and DFT calculations has led to important findings about the active sites of MoS_2 based catalysts (Jaramillo et al. 2007). One of the most significant results was the discovery of the so-called brim states and their role in HDS catalysis. This special electronic edge state can easily be seen in the STM images as a very bright brim along the edges in these MoS_2 based nanoclusters. STM analysis showed that these brim states arise from a perturbation of the electronic structure near the edges. Detailed analysis using DFT evidenced the presence of theses brim edge states, which are metallic and localized at the cluster edges (Besenbacher et al. 2008). Hydrogenation reaction is suggested to occur around the brim. Therefore, insight into these states is essential for understanding hydrotreating reactions.

Lauritsen et al. (2007) investigated the atomic-scale structure and morphology of individual Co–Mo–S and Ni–Mo–S nanoclusters synthesized on a gold substrate, which can be seen as model systems for Co- and Ni-promoted MoS_2 hydrotreating catalysts. In accordance with the widely accepted Co–Mo–S model, they observed a distinct tendency of Co and Ni substituting Mo atoms at edge sites of single-layer MoS_2 nanoclusters, which leads to truncation of the cluster morphology from the triangle of unpromoted MoS_2 to hexagon of promoted cluster. An analysis of atom-resolved STM images showed that the substitution occurred only at very specific S edge sites in Co–Mo–S and Ni–Mo–S.

Thiophene (C_4H_4S) is a main sulfur component in gasoline fuel. It represents a typical probe molecule, and has been extensively used to investigate the reactivity and kinetics of HDS reactions. It is the simplest S-containing molecule with a stabilizing aromatic system. Using the room temperature STM technique to investigate the reaction pathways of thiophene HDS, it is confirmed that interactions mainly occur at the edge domains. And both sulfur vacancies on the edges and the flat-lying molecules on the "brim" sites are important (Lauritsen, 2003 and 2004). With identification of S σ-type bonding and η5-type bonding of aromatic component, it is tentatively suggested that the desulfurization and the hydrogenation of thiophene occurs at the edge and brim sites, respectively.

Recently, Tuxen et al. (2012) studied the HDS reactions of DBT on MoS_2 and CoMoS clusters. It was observed that the accessible corner sites exist on the smallest MoS_2 nanoparticles which can strongly adsorb the S in DBT. The reactive corner sites were also observed on the hexagonal CoMoS nanoparticles. Despite the less apparent accessibility for adsorption of DBT on the Co substituted corner sites due to the hindrance of aromatics around both sides, it is indeed observed that DBT adsorbs strongly on these sites. This may be part of the reason for the strong promoting role of Co for deep desulfurization.

The low activity of the DDS pathway for 4,6-DMDBT in the industrial catalyst have been the main challenge in diesel ultra-deep HDS. The study of Grønborg et al. (2016) shown that the only option for direct chemisorption on the CoMoS nanocluster of this molecule is through a vacancy in the corner position of the S-edge in an σ-adsorption. However DFT shows that this vacancy formation is unfavorable under HDS conditions and is present only at a very low frequency which explains the low catalytic activity. Interestingly, it is found that the σ-adsorption can transform reversibly into a π-like adsorption which was observed by STM movies. This dynamic behavior is highly interesting as it reveals the possibility to shift between the two configurations and hence to alternate between the two proposed active sites. In this way both hydrogenation and direct desulfurization may happen in the same adsorption process but in different configurations around the same site, namely, on-top brim π-adsorption for hydrogenation and the σ-chemisorption in cluster corner for direct desulfurization.

Understanding the states of catalysts in real reaction conditions is beneficial for the application of industrial catalysis. It has long been believed that the stabilized MoS_2 based catalyst (after HDS) may present a new "carbide"

The Fundamental Research and Application Progress

Figure 2. Replacement of sulfur atoms by carbon atoms on the S edge with 100% sulfur coverage and 100% Mo atoms substituted by Co promoters. Simultaneous sulfidation and carburization by CH_3S or CH_3SSCH_3 on the S edge and the Mo edge.
Source: Adapted with permission from Ge et al. (2015). Copyright of American Chemical Society.

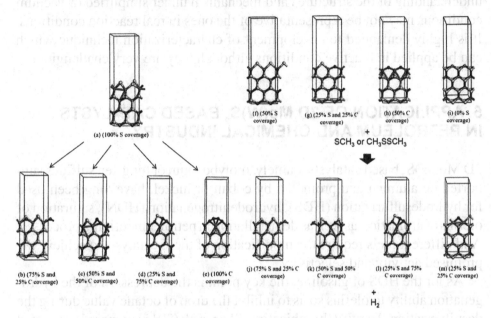

phase "Co-Mo-C" analogs to Co-Mo-S (Brysse et al. 2003). It was also evidenced (Berhault et al. 2002) that the cobalt can substantially facilitate the carbon replacement of sulfur atoms at the edges of MoS_2 layers. This is supported by DFT calculation, Wen et al. (2006) showed that replacement of sulfur atom with carbon atom at the edges of MoS_x clusters is energetically favorable. However, STM experiment combined with DFT research (Tuxen et al. 2011) suggested that incorporation of carbon in MoS_2-based catalysts as carbide type phase is not favorable when synthesized with or exposed to dimethyldisulfide (DMDS) or DMS. Recently, the new evidence of DFT (Ge et al. 2015) shows that only simultaneous carbonization and sulfidation is energetic favorable. Carbonization on pre-sulfided edge is thermodynamic prohibition (Figure 2).

The STM characterization experiment and DFT calculation have provided the key fundamental insight into surface reactions and highlighted the role of active sites in catalysis. The present progress elucidates the important role of cluster morphologies, size, doping and defects in MoS_2-based catalysts. In particular, the atomic-scale insight into the active sites in the promoted CoMoS model catalyts has led to very detailed knowledge on the nature and locations of the catalytic reaction, and has been used to develop better hydrotreating industrial catalysts.

Due to the complex and diversity of Mo(W)S_2 based catalyst, the obtained understanding of the structures and mechanism under simplified or vacuum conditions may not be representative of the ones in real reaction conditions. It is highly demanded to development of characterization technique which can be applied in reactive conditions although they are very challenging.

5. APPLICATION OF 2D Mo(W)S_2 BASED CATALYSTS IN PETROLEUM AND CHEMICAL INDUSTRY

2D Mo(W)S_2 based catalysts, namely, molybdenum or tungsten sulfides supported on alumina and promoted by cobalt or nickel, have long been used for hydrodesulfurization (HDS), hydrodenitrogenation (HDN), saturation of olefins or aromatics, as well as demetallation in petroleum refining processes. And different fuels require the modification of these catalysts to achieve the promised activity and selectivity.

As for the HDS of gasoline, the key point is the depression of the hydrogenation ability to olefins so as to inhibit the drop of octane value during the desulfurization. To fulfil this objective, Shan et al. (2015) prepared supported NiW catalysts with tunable size and morphology of active phases which can achieve highly selective HDS performance in hydro-upgrading fluid catalytic cracking (FCC) naphtha. This catalyst features W-based hybrid nanocrystals (HNCs) as the W precursor. The W-based HNCs are monodispersed ~2 nm particles on support and show a core-shell structure with $W_6O_{19}^{2-}$ as the inorganic core and short-chain quaternary ammonium cations as the organic shell. It was found that the use of W-based HNCs facilitates the formation of highly dispersed WS_2 active phases with enhanced stacking, thus yielding a larger number of accessible Ni-W-S edge sites upon nickel impregnation. Moreover, the highly W dispersion in the HNC-derived monometallic catalyst precursor allows further modification of the active phase via the optimized incorporation of Ni promoter, leading to the formation of Ni-W-S edge sites with a lower percentage of brim sites and thereby resulting in selective HDS performance for FCC naphtha. This novel approach makes it easy to tune the size and morphology of WS_2 nanoparticles, shedding light on the rational design of supported WS_2 catalysts for FCC naphtha HDS with low olefins hydrogenation.

Nikulshin et al. (2014) reported the structure-activity correlations in CoMo/Al_2O_3 catalysts for selective HDS and hydrogenation of oilfins (HYDO) of

FCC gasoline. The HDS/HYDO selectivity was found linearly dependence on the number of CoMo sites on the edges. Moreover, it was revealed that increase of the average slab length of active phase crystallites as well as (Co/Mo) edge ratio led to the increase of HDS/HYDO selectivity. Li et al. (2016) also reported that the HDS/HYDO selectivity of model gasoline can be apparently improved with the Co cooperated into the unsupported $MoS_{2\pm x}$ catalysts. When the S/Mo atomic ratio in $Co/MoS_{2\pm x}$ catalysts is increased, both HDS and HYDO activities are facilitated, however the HDS/HYDO is dropped.

Ishutenko et al. (2014) recently modified the active centers of hydrogenation (HYD) reactions by potassium in order to improve HDS/HYDO selectivity. The catalytic properties of synthesized K_x-MoP/Al_2O_3 and K_x-CoMoP/Al_2O_3 catalysts were greatly affected by potassium in hydrotreatment of a mixture of thiophene and n-hexene-1. Both thiophene and n-hexene-1 conversions decreased with the rise of K modifier content for unpromoted catalysts as well as for Co-promoted ones. Moreover, for both series of the catalysts, HYD reactions were more sensible to alkali modifier addition than HDS. As a result, the HDS/HYDO selectivity factor increased with potassium incorporation. It is supposed that potassium partially inserts in sulfide slab and probably forms of a new type of active KCoMoS sites. In the FCC gasoline hydrotreatment, the selectivity factor of HDS/HYDO increased with increasing edge-to corner ratio of active particles as well as the amount of KCoMoS species in the catalysts (Nikulshin 2016).

Recently, the sulfur content of diesel fuel have been required to reduce to ultra-low levels (10-15 ppm) by environmental regulations in many countries around the world with the intention of lowering diesel engine's harmful exhaust emissions and improving air quality. Sulfur compounds that contain alkyl side chains in the 4- and 6-positions in the dibenzothiophene molecule close to the sulfur atom (e.g. 4,6-dimethyl dibenzothiophene (4,6-DMDBT); 4-methyl,6-ethyl dibenzothiophene) are difficult to desulfurize under conventional desulfurization conditions. Hydrogenation to aromatics cycles can decrease the steric hindrance to sulfur and facilitate the desulfurization. This enhancement of simultaneously hydrogenation and desulfurization become the main manner for the improvement of $Mo(W)S_2$ based catalysts used for the ultra-deep HDS of diesel.

The HDS of diesel is usually via two paths, namely the direct desulfurization (DDS) or hydrogenation (HYD) of aromatic cycle and then desulfurization (Figure 3). Design catalyst for ultra-deep hydrodesulfurization of diesel should focus on how to remove 4,6-DMDBT more effectively (Stanishlaus 2010). These key points can be noted when modifying catalyst formulations: 1) enhance hydrogenation of aromatic ring in 4,6-DMDBT by increasing

Figure 3. The reaction pathways for HDS of DBTs
Source: Reprinted with permission from Gutiérrez et al. (2014). Copyright of American Chemical Society.

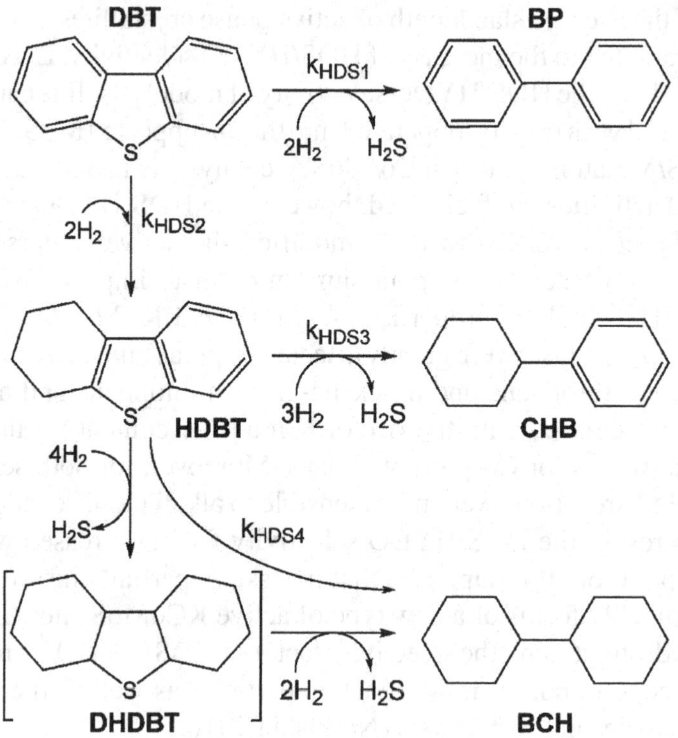

hydrogenating ability of the catalyst; 2) incorporate acidic feature in catalyst to induce isomerization of methyl groups away from the 4- and 6-positions; 3) remove inhibiting substances (such as H_2S and nitrogen containing compounds) and tailoring the reaction conditions.

Conventional Al_2O_3-supported CoMo and NiMo catalysts could not meet the demand of deep HDS to achieve 10-15 ppm sulfur in diesel products, even at higher reaction temperatures of 340-360 °C. Azizi et al. (2013) shown that high acidity alumina coated zeolite (ACZ)-supported catalysts exhibits higher HDS activity compared to Al_2O_3-supported catalysts. However, they shows high cracking activity and coke formation, which make them unsuitable for deep HDS of diesel faction at high temperatures. Low-acidic, non-polar (TiO_2 and powdered activated carbon (PAC)) supported catalysts could achieve the deep HDS of light cycle oil (LCO) at 350-360 °C. The HDS activity of TiO_2-based catalysts was better than Al_2O_3-supported catalysts and their cracking activity was significantly lower than Al_2O_3- and ACZ-supported catalysts. And the CoMo/PAC resulted in reduction of total aromatics by about 50%

leading to a significant cetane number improvement. They concluded that low-acidic, nonpolar supports (such as TiO_2 and PAC) are more suitable for CoMo and NiMo catalysts to achieve deep HDS of aromatic-rich feedstock such as LCO.

P-loaded ternary NiMoW/γ-Al_2O_3 sulfide catalyst was demonstrated having excellent hydrotreating activity in the HDS and HDN of coker light gas oil than the bimetallic NiMo(W)/Al_2O_3 or commercial catalysts (Trejo et al. 2014). The enhancement in HDN activity with phosphorous addition was attributed more to the effect of acidity than to the improvement in dispersion.

Chelating agent assisted method has been investigated and attained some industrial application in the preparation of HDS catalysts recently. For example, a series of NiMo catalysts supported on 3D cubic mesoporous KIT-6 silica were prepared by co-impregnation method using citric acid (CA) as chelating agent to adjust the sulfidation and dispersion degrees of active phase (Wu et al 2014). The addition of CA results in the formation of $MoO_3(H_2cit)^-$ and $Ni(Hcit)(cit)^{3-}$ complexes, which favors the formation of well dispersed active Ni-Mo-S phase. It was found that the addition of CA exhibited a promotion effect on HYD route in the HDS of 4,6-DMDBT. NiMoKC2 catalyst with the molar ratio of CA/Ni = 2 exhibited the highest reaction rate and the highest HYD selectivity.

Reuse of spent catalysts is beneficial to cost control. A post-treatment method in which the chelating agents are added into the calcined oxidic Mo(W) based catalysts has been tried to regenerate spent HDS catalysts (Dufresne 2007, McCarthy et al 2011). Bui et al. regenerated industrial CoMo/Al_2O_3 catalysts using maleic acid (MA) as the re-dispersant and found MA could not only re-disperse Anderson molybdenum salt species, but also promote the undesired Co species, e.g. $CoMoO_4$ to form Co-MA complex and delay their sulfidation, thus favoring the formation of Co-Mo-S active species.

Molybdenum sulfide based catalysts are also among the most promising one for higher alcohol synthesis from syngas due to the high selectivity to linear alcohols, slow deactivation, and low sensitivity to CO_2 (Subramani and Gangwal 2008). The hydrogenation of carbon monoxide to the desired oxygenates takes place thermochemical at high pressures and intermediate temperatures (Christensen et al. 2009). This investigation began in the 1980s when researchers from Dow Chemical and Union Carbide reported the ability of alkali promoted MoS_2 to produce C1- C5 alcohols from syngas. The selectivity to alcohols follows the Anderson-Schultz-Flory distribution normally, thus higher alcohol formation is limited. Doping with transition metals was found to be beneficial for the selective synthesis of higher alcohols. The addition of Co, Ni, Rh and Pd are reported to improve catalytic activity and selectivity for C2+ alcohols over MoS_2-based catalysts (Li et al. 2000, Iranmahboob et al. 2003 and Li et al. 2005). By co-modification of K/MoS_2 with Ni and Mn, the conversion of feed and the ratio of C2–C3 alcohols in products were enhanced owing to the synergistic effect of both promoters. Ni was suggested to enhance the C1→C2 homologation step (Qi et al. 2003). K- and Co-promoted MoS_2 catalysts were reported to effectively produce alcohol from syngas, with alcohol selectivity higher than 75% and total alcohol yield as high as 400 mg gcat^{-1} h^{-1} (Surisetty et al. 2012).

However, addition of H_2S is usually required to maintain the activity and higher alcohol selectivity of MoS_2-based catalysts, which results in about 2 wt% sulfur in the condensed alcohol products (Christensen et al. 2009). This renders this catalytic system problematic, especially for fuel product applications with strict sulfur specifications. The soar syngas which already contains H_2S may be more fitful for high alcohol synthesis for the MoS_2-based catalysts. The S impurities can be removed after the higher alcohols synthesis.

Improvement of $Mo(W)S_2$ based catalysts for industrial application is a complicated process. The catalysts usually contains two main parts, namely, support and active metals. A general classification of a catalyst components and their key parameters is shown in Figure 4.

Alumina is the most widely used support, but many other supports based on mixed oxides and zeolites have been investigated in recent years. The catalytic materials formulations can be improved by using different supports

Figure 4. Schematic improvement of MoS_2 based HDS catalysts

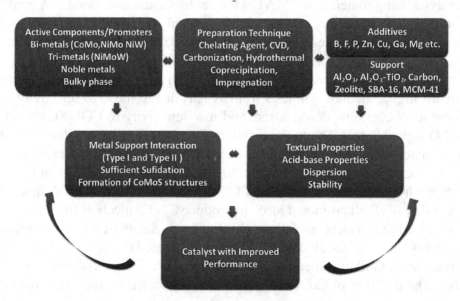

(carbon, TiO_2, TiO_2-Al_2O_3, HY, MCM-41, etc.). A balance between hydrogenation of metal sites and the cracking of support is critical in order to synthesize a bifunctional catalyst for hydrotreatment of heavy fuels. To alumina-supported CoMo, NiMo and NiW catalysts aiming to diesel deep HDS, the improvement can be achieved by increasing loading of active metal (Mo, W, etc.); by adding one more transient metal (e.g. Ni to CoMo or Co to NiMo); and by incorporating a noble metal (Pt, Pd, Ru, etc.), but these methods may accordingly lead to the increase of cost. For gasoline hydrotreatment, a precondition is high selectivity to desulfurization with low hydrogenation of olefins, but this is still challenging (Saleh 2016). In future, development of synthesizing methodologies which can sufficiently utilize the active atoms, decrease the cost and achieve high activity and selectivity will be the ultimate objectives. These must be based on the deep understanding of the reaction mechanism, active structures and process engineering. And this is becoming possible with the fast advances in fundamental research and industrial application.

REFERENCES

Azizi, N., Ali, S. A., Alhooshani, K., Kim, T., Lee, Y., Park, J.-I., & Mochida, I. et al. (2013). Hydrotreating of light cycle oil over NiMo and CoMo catalysts with different supports. *Fuel Processing Technology*, *109*, 172–178. doi:10.1016/j.fuproc.2012.11.001

Berhault, G., Cota Araiza, L., Duarte Moller, A., Mehta, A., & Chianelli, R. R. (2002). Modifications of Unpromoted and Cobalt-Promoted MoS_2 during Thermal Treatment by Dimethylsulfide. *Catalysis Letters*, *78*(1), 81–90. doi:10.1023/A:1014910105975

Besenbacher, F., Brorson, M., Clausen, B. S., Helveg, S., Hinnemann, B., Kibsgaard, J., & Topsøe, H. et al. (2008). Recent STM, DFT and HAADF-STEM studies of sulfide-based hydrotreating catalysts: Insight into mechanistic, structural and particle size effects. *Catalysis Today*, *130*(1), 86–96. doi:10.1016/j.cattod.2007.08.009

Breysse, M., Afanasiev, P., Geantet, C., & Vrinat, M. (2003). Overview of support effects in hydrotreating catalysts. *Catalysis Today*, *86*(1–4), 5–16. doi:10.1016/S0920-5861(03)00400-0

Christensen, J. M., Mortensen, P. M., Trane, R., Jensen, P. A., & Jensen, A. D. (2009). Effects of H2S and process conditions in the synthesis of mixed alcohols from syngas over alkali promoted cobalt-molybdenum sulfide. *Applied Catalysis A, General*, *366*(1), 29–43. doi:10.1016/j.apcata.2009.06.034

Delmon, B., & Froment, G. F. (1996). Remote Control of Catalytic Sites by Spillover Species: A Chemical Reaction Engineering Approach. *Catalysis Reviews*, *38*(1), 69–100. doi:10.1080/01614949608006454

Dufresne, P. (2007). Hydroprocessing catalysts regeneration and recycling. *Applied Catalysis A, General*, *322*, 67–75. doi:10.1016/j.apcata.2007.01.013

Fan, J., Yu, C., Gao, F., Lei, J., Tian, B., Wang, L., ... Zhao, D. (2003). Cubic Mesoporous Silica with Large Controllable Entrance Sizes and Advanced Adsorption Properties. Angewandte Chemie International Edition, 42(27), 3146–3150.

Ge, H., Wen, X.-D., Ramos, M. A., Chianelli, R. R., Wang, S., Wane, J., & Li, X. et al. (2014). Carbonization of Ethylenediamine Coimpregnated CoMo/Al_2O_3 Catalysts Sulfided by Organic Sulfiding Agent. *ACS Catalysis*, *4*(8), 2556–2565. doi:10.1021/cs500477x

Grønborg, S. S., Šarić, M., Moses, P. G., Rossmeisl, J., & Lauritsen, J. V. (2016). Atomic scale analysis of sterical effects in the adsorption of 4,6-dimethyldibenzothiophene on a CoMoS hydrotreating catalyst. *Journal of Catalysis*, *344*, 121–128. doi:10.1016/j.jcat.2016.09.004

Gutiérrez, O. Y., Singh, S., Schachtl, E., Kim, J., Kondratieva, E., Hein, J., & Lercher, J. A. (2014). Effects of the Support on the Performance and Promotion of (Ni)MoS_2 Catalysts for Simultaneous Hydrodenitrogenation and Hydrodesulfurization. *ACS Catalysis*, *4*(5), 1487–1499. doi:10.1021/cs500034d

Guzmán, M., Huirache-Acuña, R., Coricera, C. V., Hernández, J. R., Díaz de León, J. N., & Pawelec, B. (2013). Removal of refractory S-containing compounds from liquid fuels over P-loaded NiMoW/SBA-16 sulfide catalysts. *Fuel*, *103*, 321–333. doi:10.1016/j.fuel.2012.07.005

Harris, S. (1986). Catalysis by transition metal sulfides: A theoretical and experimental study of the relation between the synergic systems and the binary transition metal sulfides. *Journal of Catalysis*, *98*(1), 17–31. doi:10.1016/0021-9517(86)90292-7

Huirache-Acuña, R., Pawelec, B., Loricera, C. V., Rivera-Muñoz, E., Nava, R., Torres, B., & Fierro, J. L. G. (2012). Comparison of the morphology and HDS activity of ternary Ni(Co)-Mo-W catalysts supported on Al-HMS and Al-SBA-16 substrates. *Applied Catalysis B: Environmental, 125*, 473–485. doi:10.1016/j.apcatb.2012.05.034

Huirache-Acuña, R., Pawelec, B., Rivera-Muñoz, E., Nava, R., Espino, J., & Fierro, J. L. G. (2009). Comparison of the morphology and HDS activity of ternary Co-Mo-W catalysts supported on P-modified SBA-15 and SBA-16 substrates. *Applied Catalysis B: Environmental, 92*(1–2), 168–184. doi:10.1016/j.apcatb.2009.07.012

Iranmahboob, J., Hill, D. O., & Toghiani, H. (2002). K_2CO_3/Co-MoS_2/clay catalyst for synthesis of alcohol: Influence of potassium and cobalt. *Applied Catalysis A, General, 231*(1–2), 99–108. doi:10.1016/S0926-860X(01)01011-0

Ishutenko, D., Nikulshin, P., & Pimerzin, A. (2016). Relation between composition and morphology of K(Co)MoS active phase species and their performances in hydrotreating of model FCC gasoline. *Catalysis Today, 271*, 16–27. doi:10.1016/j.cattod.2015.11.025

Jaramillo, T. F., Jørgensen, K. P., Bonde, J., Nielsen, J. H., Horch, S., & Chorkendorff, I. (2007). Identification of active edge sites for electrochemical H2 evolution from MoS_2 nanocatalysts. *Science, 317*(5834), 100–102. doi:10.1126/science.1141483 PMID:17615351

Kibsgaard, J., Tuxen, A., Knudsen, K. G., Brorson, M., Topsøe, H., Lægsgaard, E., & Besenbacher, F. et al. (2010). Comparative atomic-scale analysis of promotional effects by late 3d-transition metals in MoS_2 hydrotreating catalysts. *Journal of Catalysis, 272*(2), 195–203. doi:10.1016/j.jcat.2010.03.018

Laurenti, D., Phung-Ngoc, B., Roukoss, C., Devers, E., Marchand, K., Massin, L., & Vrinat, M. et al. (2013). Intrinsic potential of alumina-supported CoMo catalysts in HDS: Comparison between γc, γT, and δ-alumina. *Journal of Catalysis, 297*, 165–175. doi:10.1016/j.jcat.2012.10.006

Lauritsen, J. V., Kibsgaard, J., Olesen, G. H., Moses, P. G., Hinnemann, B., Helveg, S., & Besenbacher, F. et al. (2007). Location and coordination of promoter atoms in Co- and Ni-promoted MoS_2-based hydrotreating catalysts. *Journal of Catalysis, 249*(2), 220–233. doi:10.1016/j.jcat.2007.04.013

Lauritsen, J. V., Nyberg, M., Nørskov, J. K., Clausen, B. S., Topsøe, H., Lægsgaard, E., & Besenbacher, F. (2004). Hydrodesulfurization reaction pathways on MoS$_2$ nanoclusters revealed by scanning tunneling microscopy. *Journal of Catalysis, 224*(1), 94–106. doi:10.1016/j.jcat.2004.02.009

Lauritsen, J. V., Nyberg, M., Vang, R. T., Bollinger, M. V., Clausen, B. S., Topsøe, H., & Besenbacher, F. et al. (2003). Chemistry of one-dimensional metallic edge states in MoS2 nanoclusters. *Nanotechnology, 14*(3), 385–389. doi:10.1088/0957-4484/14/3/306

Lélias, M. A., Kooyman, P. J., Mariey, L., Oliviero, L., Travert, A., van Gestel, J., & Maugé, F. et al. (2009). Effect of NTA addition on the structure and activity of the active phase of cobalt-molybdenum sulfide hydrotreating catalysts. *Journal of Catalysis, 267*(1), 14–23. doi:10.1016/j.jcat.2009.07.006

Li, D., Yang, C., Li, W., Sun, Y., & Zhong, B. (2005). Ni/ADM: A high activity and selectivity to C2+OH catalyst for catalytic conversion of synthesis gas to C1-C5 mixed alcohols. *Topics in Catalysis, 32*(3–4), 233–239. doi:10.1007/s11244-005-2901-x

Li, P., Liu, X., Zhang, C., Chen, Y., Huang, B., Liu, T., & Li, C. et al. (2016). Selective hydrodesulfurization of gasoline on Co/MoS2±x catalyst: Effect of sulfur defects in MoS2±x. *Applied Catalysis A, General, 524*, 66–76. doi:10.1016/j.apcata.2016.06.003

Li, Y. W., & Delmon, B. (1997). Modelling of hydrotreating catalysis based on the remote control: HYD and HDS. *Journal of Molecular Catalysis A Chemical, 127*(1–3), 163–190. doi:10.1016/S1381-1169(97)00121-0

Li, Z., Fu, Y., Jiang, M., Meng, M., Xie, Y., Hu, T., & Liu, T. (2000). Structures and performance of Pd–Mo–K/Al$_2$O$_3$ catalysts used for mixed alcohol synthesis from synthesis gas. *Catalysis Letters, 65*(1), 43–48. doi:10.1023/A:1019017321625

Liu, C., Yuan, P., & Cui, C. (2016a). The Pore Confinement Effect of FDU-12 Mesochannels on MoS$_2$ Active Phases and Their Hydrodesulfurization Performance. *Journal of Nanomaterials, 2016*, 1–10.

Liu, H., Li, Y., Yin, C., Wu, Y., Chai, Y., Dong, D., & Liu, C. et al. (2016b). One-pot synthesis of ordered mesoporous NiMo-Al$_2$O$_3$catalysts fordibenzo-thiophene hydrodesulfurization. *Applied Catalysis B: Environmental, 198*, 493–507. doi:10.1016/j.apcatb.2016.06.004

Liu, H., Liu, C., Yin, C., Chai, Y., Li, Y., Liu, D., & Li, X. et al. (2015). Preparation of highly active unsupported nickel–zinc–molybdenum catalysts for the hydrodesulfurization of dibenzothiophene. *Applied Catalysis B: Environmental, 174-175*, 264–276. doi:10.1016/j.apcatb.2015.02.009

Liu, H., Yin, C., Li, H., Liu, B., Li, X., Chai, Y., & Liu, C. et al. (2014). Synthesis, characterization and hydrodesulfurization properties of nickel–copper–molybdenum catalysts for the production of ultra-low sulfur diesel. *Fuel, 129*, 138–146. doi:10.1016/j.fuel.2014.03.055

Marchal, C., Uzio, D., Merdrignac, I., Barré, L., & Geantet, C. (2012). Study of the role of the catalyst and operating conditions on the sediments formation during deep hydroconversion of vacuum residue. *Applied Catalysis A, General, 411-412*, 35–43. doi:10.1016/j.apcata.2011.10.018

Mazoyer, P., Geantet, C., Diehl, F., Loridant, S., & Lacroix, A. (2008). Role of chelating agent on the oxidic state of hydrotreating catalysts. *Catalysis Today, 130*(1), 75–79. doi:10.1016/j.cattod.2007.07.013

McCarthy, S. J., Bai, C., Borghard, W. G., & Lewis, W. E. (2011). *US Patent No. 7906447*. Washington, DC: US Patent Office.

Mendoza-Nieto, J. A., Puente-Lee, I., Salcedo-Luna, C., & Klimova, T. (2012). Effect of titania grafting on behavior of NiMo hydrodesulfurization catalysts supported on different types of silica. *Fuel, 100*, 100–109. doi:10.1016/j.fuel.2012.02.005

Mozhaev, A., & Nilulshin, P. (2016). Investigation of co-promotion effect in NiCoMoS/Al_2O_3 catalysts based on Co2Mo10-heteropolyacid and nickel citrate. *Catalysis Today, 271*, 80–90. doi:10.1016/j.cattod.2015.11.002

Nikulshin, P. A., Ishutenko, D. I., Mozhaev, A. A., Maslakov, K. I., & Pimerzin, A. A. (2014). Effects of composition and morphology of active phase of CoMo/Al_2O_3 catalysts prepared using Co2Mo10–heteropolyacid and chelating agents on their catalytic properties in HDS and HYD reactions. *Journal of Catalysis, 312*, 152–169. doi:10.1016/j.jcat.2014.01.014

Nilulshin, P., Ishutenko, D., Anashkin, Y., Mozhaev, A., & Pimerzin, A. (2016). Selective hydrotreating of FCC gasoline over KCoMoP/Al_2O_3 catalysts prepared with H3PMo12O40: Effect of metal loading. *Fuel, 182*, 632–639. doi:10.1016/j.fuel.2016.06.016

Ojeda, J., Escalona, N., Baeza, P., Escudey, M., & Gil-Llambías, F. J. (2003). Synergy between Mo/SiO$_2$ and Co/SiO$_2$ beds in HDS: A remote control effect? *Chemical Communications (Cambridge)*, (13): 1608–1609. doi:10.1039/B301647C

Okamoto, Y., Hioka, K., Arakawa, K., Fujikawa, T., Ebihara, T., & Kubota, T. (2009). Effect of sulfidation atmosphere on the hydrodesulfurization activity of SiO$_2$-supported Co–Mo sulfide catalysts: Local structure and intrinsic activity of the active sites. *Journal of Catalysis*, *268*(1), 49–59. doi:10.1016/j.jcat.2009.08.017

Pimerzin, A. A., Nikulshin, P. A., Mozhaev, A. V., Pimerzin, A. A., & Lyashenko, A. I. (2015). Investigation of spillover effect in hydrotreating catalysts based on Co2Mo10− heteropolyanion and cobalt sulphide species. *Applied Catalysis B: Environmental*, *168–169*, 396–407. doi:10.1016/j.apcatb.2014.12.031

Qi, H. (2003). Nickel and manganese co-modified K/MoS2 catalyst: High performance for higher alcohols synthesis from CO hydrogenation. *Catalysis Communications*, *4*(7), 339–342. doi:10.1016/S1566-7367(03)00061-X

Ramos, M., Berhault, G., Ferrer, D. A., Torres, B., & Chianelli, R. R. (2012). HRTEM and molecular modeling of the MoS$_2$–Co$_9$S$_8$ interface: Understanding the promotion effect in bulk HDS catalysts. *Catal. Sci. Technol.*, *2*(1), 164–178. doi:10.1039/C1CY00126D

Rinaldi, N., Kubota, T., & Okamoto, Y. (2009). Effect of Citric Acid Addition on Co-Mo/B2O3/Al2O3 Catalysts Prepared by a Post-Treatment Method. *Industrial & Engineering Chemistry Research*, *48*(23), 10414–10424. doi:10.1021/ie9008343

Rinaldi, N., Kubota, T., & Okamoto, Y. (2010). Effect of citric acid addition on the hydrodesulfurization activity of MoO$_3$/Al$_2$O$_3$ catalysts. *Applied Catalysis A, General*, *374*(1–2), 228–236. doi:10.1016/j.apcata.2009.12.015

Saleh, T. A. (2016). *Applying Nanotechnology to the Desulfurization Process in Petroleum Engineering*. IGI Global. Retrieved from http://services.igi-global.com/resolvedoi/resolve.aspx?doi=10.4018/978-1-4666-9545-0

Shan, S., Yuan, P., Han, W., Shi, G., & Bao, X. (2015). Supported NiW catalysts with tunable size and morphology of active phases for highly selective hydrodesulfurization of fluid catalytic cracking naphtha. *Journal of Catalysis*, *330*, 288–301. doi:10.1016/j.jcat.2015.06.019

Stanishlaus, A., Marafi, A., & Rana, M. S. (2010). Recent advances in the science and technology of ultra low sulfur diesel (ULSD) production. *Catalysis Today*, *153*(1-2), 1–68. doi:10.1016/j.cattod.2010.05.011

Subramani, V., & Gangwal, S. K. (2008). A Review of Recent Literature to Search for an Efficient Catalytic Process for the Conversion of Syngas to Ethanol. *Energy & Fuels*, *22*(2), 814–839. doi:10.1021/ef700411x

Sun, M., Nicosia, D., & Prins, R. (2003). The effects of fluorine, phosphate and chelating agents on hydrotreating catalysts and catalysis. *Catalysis Today*, *86*(1–4), 173–189. doi:10.1016/S0920-5861(03)00410-3

Surisetty, V. R., Dalai, A. K., & Kozinski, J. (2011). Alcohols as alternative fuels: An overview. *Applied Catalysis A, General*, *404*, 1–11.

Surisetty, V. R., Eswaramoorthi, I., & Dalai, A. K. (2012). Comparative study of higher alcohols synthesis over alumina and activated carbon-supported alkali-modified MoS_2 catalysts promoted with group VIII metals. *Fuel*, *96*, 77–84. doi:10.1016/j.fuel.2011.12.054

Surisetty, V. R., Tavasoli, A., & Dalai, A. K. (2009). Synthesis of higher alcohols from syngas over alkali promoted MoS2 catalysts supported on multi-walled carbon nanotubes. *Applied Catalysis A, General*, *365*(2), 243–251. doi:10.1016/j.apcata.2009.06.017

Topsøe, H. (1981). In situ Mössbauer emission spectroscopy studies of unsupported and supported sulfided Co-Mo hydrodesulfurization catalysts: Evidence for and nature of a Co-Mo-S phase. *Journal of Catalysis*, *68*(2), 433–452. doi:10.1016/0021-9517(81)90114-7

Topsøe, H. (2007). The role of Co–Mo–S type structures in hydrotreating catalysts. *Applied Catalysis A, General*, *322*, 3–8. doi:10.1016/j.apcata.2007.01.002

Trejo, F., Rana, M. S., Ancheta, J., & Chávez, S. (2014). Influence of support and supported phases on catalytic functionalities of hydrotreating catalysts. *Fuel*, *138*, 104–110. doi:10.1016/j.fuel.2014.02.032

Tuxen, A., Gøbel, H., Hinnemann, B., Li, Z., Knudsen, K. G., Topsøe, H., & Besenbacher, F. et al. (2011). An atomic-scale investigation of carbon in MoS_2 hydrotreating catalysts sulfided by organosulfur compounds. *Journal of Catalysis*, *281*(2), 345–351. doi:10.1016/j.jcat.2011.05.018

Tuxen, A. K., Füchtbauer, H. G., Temel, B., Hinnemann, B., Topsøe, H., Knudsen, K. G., & Lauritsen, J. V. et al. (2012). Atomic-scale insight into adsorption of sterically hindered dibenzothiophenes on MoS_2 and Co–Mo–S hydrotreating catalysts. *Journal of Catalysis, 295*, 146–154. doi:10.1016/j.jcat.2012.08.004

Vatutina, Y. V., Klimov, O. V., Nadeina, K. A., Danilova, I. G., Gerasimov, E. Y., Prosvirin, I. P., & Noskov, A. S. (2016). Influence of boron addition to alumina support by kneading on morphology and activity of HDS catalysts. *Applied Catalysis B: Environmental, 199*, 23–32. doi:10.1016/j.apcatb.2016.06.018

Villarroel, M., Baeza, P., Escalona, N., Ojeda, J., Delmon, B., & Gil-Llambías, F. J. (2008). MD//Mo and MD//W [MD=Mn, Fe, Co, Ni, Cu and Zn] promotion via spillover hydrogen in hydrodesulfurization. *Applied Catalysis A, General, 345*(2), 152–157. doi:10.1016/j.apcata.2008.04.033

Walton, A. S., Lauritsen, J. V., Topsøe, H., & Besenbacher, F. (2013). MoS_2 nanoparticle morphologies in hydrodesulfurization catalysis studied by scanning tunneling microscopy. *Journal of Catalysis, 308*, 306–318. doi:10.1016/j.jcat.2013.08.017

Wang, W., Li, L., Tan, S., Wu, K., Zhu, G., Liu, Y., & Yang, Y. et al. (2016). Preparation of NiS2//MoS2 catalysts by two-step hydrothermal method and their enhanced activity for hydrodeoxygenation of p-cresol. *Fuel, 179*, 1–9. doi:10.1016/j.fuel.2016.03.068

Wen, X.-D., Cao, Z., Li, Y.-W., Wang, J., & Jiao, H. (2006). Structure and Energy of $Mo_{27}S_xC_y$ Clusters: A Density Functional Theory Study. *The Journal of Physical Chemistry B, 110*(47), 23860–23869. doi:10.1021/jp063323b PMID:17125352

Wu, H., Duan, A., Zhao, Z., Qi, D., Li, J., Liu, B., & Zhang, X. et al. (2014). Preparation of NiMo/KIT-6 hydrodesulfurization catalysts with tunable sulfidation and dispersion degrees of active phase by addition of citric acid as chelating agent. *Fuel, 130*, 203–210. doi:10.1016/j.fuel.2014.04.038

Zhang, Y., Han, W., Long, X., & Nie, H. (2016). Redispersion effects of citric acid on CoMo/γ-Al_2O_3 hydrodesulfurization catalysts. *Catalysis Communications, 82*, 20–23. doi:10.1016/j.catcom.2016.04.012

Chapter 3
3D Catalysts of Mo(W) Carbide, Nitride, Oxide, Phosphide, and Boride

In last two chapters, we have shown the exciting catalytic potential of the 2D Mo(W) dichalcogenides. Other Mo(W) based materials, such as Mo(W) nitride, carbide, oxide and phosphide, are usually in some 3D structures, catalytic performance depends on their morphology, size and exposed crystal facet. In this chapter, we focus on the preparation, characterization and novel catalytic applications of these 3D Mo(W) materials.

1. THE SYNTHESIS AND CHARACTERIZATION OF Mo(W) CARBIDE

Mo(W) carbide with a list of desired properties makes it very attractive for catalysis from both fundamental and industrial standpoints. Mo(W)-based carbides have been found to show special catalytic properties similar to noble metals. This has resulted in surging investigation on synthesis of various Mo_2C

DOI: 10.4018/978-1-5225-2274-4.ch003

structures, from nanoparticles to nanorods (Vrubel and Hu 2012, Chen et al. 2013a and 2013b, Youn et al. 2014 and Xiao et al. 2015). The methodologies and conditions used for molybdenum (tungsten) carbide preparation determine the chemical and catalytic nature of the obtained materials (Wan et al. 2014 and Wu et al. 2015).

Traditionally, the Mo(W) carbide material is usually synthesized by temperature-programmed reaction (TPR) where a given amount of the oxide precursor (for instance, MoO_3 or WO_3) is heated gradually under a mixture of hydrogen and hydrocarbon. The oxide firstly is reduced from high valence to lower valence oxide states and then carburized at high temperature. Normally the formed molybdenum carbide is stable hexagonal phase. The formed carbide is passivated using 0.5-1% O_2 before exposing to air owing to its pyrophoric property. Using the TPR procedure, many parameters are necessary to consider such as carburizing agent, C/Mo ratio, heating rate, isothermal treatment, and H_2 concentration.

MoO_2 is deemed the common intermediate in carburization process. Guzmán et al. (2013 and 2015) synthesized small MoO_2 nanoparticles using EG as reducing agent to fabricate high-dispersed Mo carbide. In-situ time resolved X-ray diffraction shows that the most intense peak of the monoclinic MoO_2 phase (011) became broad suggesting the presence of small crystalline domains (Figure 1).

Xiao et al. (2001) compared various molybdenum carbides prepared by the temperature programmed reaction method using mixtures of hydrogen and methane, hydrogen and ethane, or hydrogen and butane. The results show that the choice of hydrocarbon used to synthesize molybdenum carbide significantly affects the structure and texture of the resultant materials. Increasing the chain length of the carburizing agent reduces the particle size and the temperature for phase transformation. Carburizing with hydrogen/methane gives rise to hexagonal closed packed (hcp) carbide (β-Mo_2C), while using butane as the carbon source, molybdenum oxide is mainly reduced to face centered cubic (fcc) carbide (α-Mo_2C). When using ethane as the carbon source, the resultant carbide has a mixed phase composition with the hcp phase dominant. With modified TPR procedure, it is possible to synthesize nanocrystalline of molybdenum carbide in either cubic or hexagonal phase (Frauwallner et al. 2011).

However, using the typical TPR method, the transformation from oxide to carbide is usually incomplete since the reactions occur only on the gas-solid interface. Moreover, the bulky products generated are often covered by carbon species originated from the pyrolysis of excessive carbon-containing gases, which suppresses the surface activity. Furthermore, the complex and

Figure 1. In situ XRD patterns of the carburization of MoO_2 under a mixture of n-heptane/H_2 showing the broadening of the carbide peaks.
Source: Reprinted with permission from Guzmán et al. (2013). Copyright of NRC Research Press.

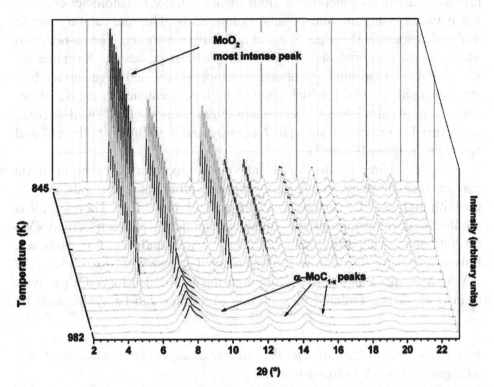

strict synthetic conditions, such as precise control of temperature ramping rate (≤1 °C/min) and gas components and flow rate, have heavily limited the large-scale manufacture of carbide materials. Therefore, easy and safe strategies for the controllable preparation of nanostructured carbides through homogeneous reactions are highly desired, especially for large-scale manufacture and application. Based on this procedure, Wan et al. (2014) synthesized four phases of molybdenum carbide (α-MoC_{1-x}, β-Mo_2C, η-MoC, and γ-MoC) using ammonium heptamolybdate and 4-Cl-o-phenylenediamine with different treatment conditions. β-Mo_2C, η-MoC, and γ-MoC have similar hexagonal crystal structures but their stacking sequences are different. β-Mo_2C shows an ABAB stacking sequence for the metal planes, while η-MoC, and γ-MoC exhibit ABCABC and AAAA packing structures respectively. The cubic α-MoC_{1-x} is isostructural with NaCl, which has an ABCABC packing. Comparative study showed that their HER activities are in the decreased sequence of β-Mo_2C > γ-MoC > η-MoC > α-MoC_{1-x}.

Gao et al. (2009) synthesized nanoporous molybdenum carbide nanowires based on MoO_x/amine hybrid (Figure 2). It was found that such hybrids could provide quasi-homogeneous reactions owing to the sub-nanometer-contacting between inorganic and organic components. The intercalating amine molecules act as both reducing agent and carbon resource. The as-obtained Mo_2C nanowires composed of small nanoparticles possess uniform one-dimensional (1D) morphology, abundant nanoporosity, and large surface free from depositing carbon, which showed high performance in methanol decomposition. MoC–Mo_2C heteronanowires composed of well-defined nanoparticles presented high HER activity and stability in both acid and basic electrolytes (Figure 3).

The surface area of Mo(W) carbides is of importance for their catalytic performance. However, applying conventional carburization methods, Mo(W) Cx with high specific surface areas are difficult to achieve. Therefore, it is required to develop appropriate synthesis techniques which can form Mo(W) Cx with nanoscale particle size, high surface area, and large pore volumes. Polyoxometalates (POMs), as a unique class of nanosized M-oxo clusters (M = W or Mo) are ideal precursors for preparing nanoscale Mo(W)Cx catalysts (Pan et al. 2014). Ordered mesoporous Mo(W)Cx can be synthesized by

Figure 2. Schematic illustration for the synthetic strategy of Mo_2C nanowires based on organic-inorganic hybrid nanowires.
Source: Reprinted with permission from Gao et al. (2009). Copyright of American Chemical Society.

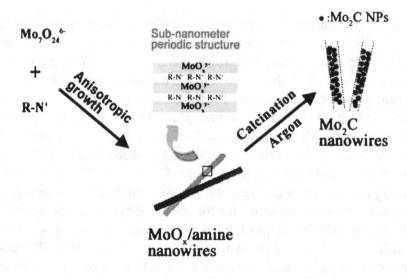

Figure 3. (a) Schematic illustration for the fabrication of MoC_x HNWs from MoO_x– amine NWs with tunable composition. (b) XRD patterns and (c) Mo 3d XPS profiles of the as-obtained MoC_x NWs.
Source: Reprinted with permission from Lin et al. (2016a).

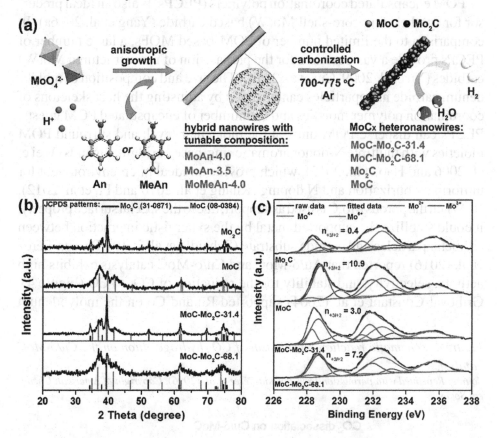

introduction of POMs into mesoporous silica followed by carburization using an additional carbon source, such as CO or CH_4 (Cui et al. 2008 and Wu et al. 2009).

Lunkenbein et al. (2012) established a direct path to synthesize mesoporous MoC/C nanocomposites by simple high-temperature (>700 °C) heat treatment of a hexagonally ordered diblock copolymer/POMs nanocomposite in an argon atmosphere. The diblock copolymer served as both template agent and carbon source. The H_3PMo units embedded in a porous carbon matrix were converted into MoO_xC_y nanoparticles. The resulting MoC/C nanocomposites exhibited inverse hexagonal order, hierarchical pore structure, and high surface area. Deng et al. (2015) and Wu et al. (2015) synthesized novel molybdenum carbides from POM-based MOFs. Liu et al. (2015) prepared hybrid materials composed of molybdenum carbide core and N-rich graphene-like carbon shells from POM precursors with organic carbon resources. All of them show remarkable HER performance.

POM encapsulated coordination polymers (PECPs) is also an ideal precursor for synthesizing core-shell Mo(W) based carbide (Yang et al. 2016a). In comparison to the limited number of POM-based MOFs, a large number of PECPs provide a vast resource for the preparation of nanostructured Mo(W) carbides (Yu et al. 2009, Cui et al. 2014). The size and composition of molybdenum carbide nanoparticles can be tuned by adjusting the host skeletons of coordination polymer moieties and the number of encapsulated POM guests. PECPs can disperse POM units at the molecular level, and surround POM moieties with multiple N-donor aromatic ring-based bridging ligands (Wei et al. 2006 and Hao et al. 2015), which provide an ideal microenvironment for uniform carburization and N doping (Kuang et al. 2010 and Fu et al. 2012).

When the porous Mo_2C is used as support, due to the special surface property, it could stabilize the supported metal by the synergistic interaction between ad-metal particles and carbide substrate (Schaidle et al. 2010). Posada-Pérez et al. (2016) reported that Au/δ-MoC and Cu/δ-MoC catalysts exhibits high activity, selectivity, and stability for the reduction of CO_2 to CO (Figure 4). Griboval-Constant et al. (2004) supported Ru and Co on the molybdenum

Figure 4. Schematic reaction mechanism of CO_2 hydrogenation on the Cu/δ-MoC catalyst.
Source: Reprinted with permission from Posada-Pérez et al. (2016). Copyright of American Chemical Society.

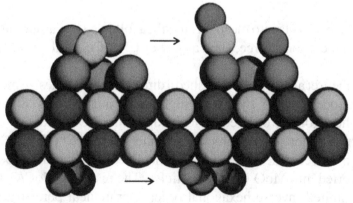

carbide for Fischer-Tropsch synthesis and evidenced an apparent promoting effect. However, it should be noted that the metal was actually supported on the passivated surface of the molybdenum carbide (i.e., oxide surface), but not on the real carbide surface.

To investigate metal impregnated carbide where the metals interact directly with the native surface of carbide; Schweitzer et al. (2011) directly deposited Pt onto the carbide surface avoiding exposure of the carbide surface to air. It revealed that Pt deposition on Mo_2C was governed primarily by the redox reactions of the Pt being reduced and the Mo being oxidized (Setthapun et al. 2008). However, this redox reaction was inhibited for the Pt/p-Mo_2C catalyst owing to the presence of the passivation layer. Thus nanoscale Pt metal particles were formed on the unpassivated Mo_2C while very large Pt particles on the passivated p-Mo_2C. The Pt/Mo_2C showed significantly higher activity of water-gas shift (WGS) compared to the Pt/p-Mo_2C catalyst (Schaidle et al. 2012). Meanwhile an excellent tolerance to sulfur was also evidenced during the steam reform of methanol (SRM) reaction (Lausche et al. 2011).

Sabnis et al. (2015a) investigated Pt deposited over passivated Mo_2C/MWCNT by XAS, EELS and HAADF-STEM techniques. As compared to the Pt foil, Pt L III XANES spectrum of 1.5wt%Pt/2wt%Mo/MWCNT after reduction at 600 °C presents a shift in the leading edge toward higher energy. However, their similar edge energies suggests that Pt is fully reduced. Thus, the observed differences are attributed to the formation of Pt–Mo bimetallic nanoparticles. The Fourier transforms of Pt L III edge EXAFS spectra shown that the Pt–Pt coordination numbers are 4.1 (after reduction) with inter-atomic distance of 2.74 Å by the parameter fittings. The Pt–Mo coordination numbers were ~4.6, suggesting the incorporation of Mo into Pt particles to form alloy. The intensity analysis using the HAADF-STEM suggested that the thickness of the alloy nanoparticles ranges from 5 to 8 layers. The synergy between the Pt–Mo alloy (CO activation) and Mo_2C (water dissociation) is proposed as the cause of the high WGS performance of Mo_2C/ MWCNT catalysts.

For decreasing the size of Mo_2C particles to expose more active sites, one can disperse them on a support. For example, Tuomi et al (2016) synthesized a Mo_2C/RGO catalyst by three sequential steps. 1) the oxidative treatment of graphene surface, 2) deposition of MoO_2 nano-particles (NPs) onto the graphene oxide (GO) surface via a supercritical alcohol route, and 3) a carbothermal hydrogen reduction (CHR) treatment transforming the MoO_2 NPs into Mo_2C NPs, companied by the conversion of GO to the reduced graphene oxide (RGO). With the carbothermal reduction method, Ma et al. (2014) synthesized Mo_2C crystalline nanoparticles on a carbonaceous support with high Mo loading.

2. THE SYNTHESIS AND CHARACTERIZATION OF Mo(W) NITRIDE

Similar to Mo(W) carbide, Mo(W) based nitrides have also attracted attention of scientists. The usual routes to synthesis of Mo(W) nitride were ammonolysis of oxide precursors by heat treatment under NH_3 atmosphere or decomposition of organometallic nitrogen containing complexes under inert gas atmosphere (Chen et al. 2012). Fan et al. (2014) investigated the influence of preparing methods of oxide precursor on activity of Ni-Mo nitride catalysts for propane ammoxidation reaction. Among the four methods of sol-gel, rotary evaporation, microwave drying, co-precipitation, impregnation and mechanical mixing, the co-precipitation one was the best in catalytic activity and selectivity. Jaggers et al. (1990) studied the influence of initial salts on the formation of binary molybdenum nitride, and reported that different precursors, such as MoO_3, $(NH_4)_6Mo_7O_{24} \cdot 4H_2O$, $(NH_4)_2MoO_4$, and H_xMoO_3, can lead to the either Mo_2N or MoN with various specific surface area values differing up to 13 times. Wang et al. (2008) suggested that the NH^{4+} in the inorganic salt can be utilized as the N source for preparation of Mo nitride.

Sometimes, nitride or carbide (including mixed species) can be produced using the same protocol only by changing preparing parameters. For example, Mo(W) nitrides and carbides were synthesized by a urea method where the urea as both a nitrogen/carbon source and a stabilizing agent for nanoparticle dispersion. The obtained molybdenum and tungsten nitride and carbides were almost pure and crystalline. Sizes are around 20 nm for Mo nitrides or carbides compared with 4 nm for W based ones. The specific surface area was 10-80 m^2/g, depending on the metal and the initial ratio of metal precursor to urea (Giordano et al. 2008 and 2009).

High-pressure synthesis have been recently demonstrated as an efficient manner in the search for new nitrides, especially the nitride-rich ones. However, preparation of nitride-rich Mo or W nitrides are deemed challenge because incorporation of nitrogen into the crystal lattice is often thermodynamically unfavorable at atmospheric pressure. As a result, most of the known Mo or W nitrides are nitrogen-deficient with $x \leq 1$ in $Mo(W)N_x$. In the binary Mo-N system, using the conventional procedure, three different phases with varying nitrogen concentrations are reported: tetragonal $\beta-MoN_x$ with $x \leq 0.5$, cubic $\gamma-MoN_x$ with $0.5 \leq x < 1$ and hexagonal $\delta-MoN_x$ with $x \geq 1$. By high pressure protocol, we successfully synthesized a series of novel N-rich tungsten nitrides (*e.g.,* W_2N_3, W_3N_4) through newly formulated ion-exchange reactions at moderate pressures up to 5 GPa (Wang et al. 2012). In addition, we extended this high-P methodology to molybdenum mononitrides and have

3D Catalysts of Mo(W) Carbide, Nitride, Oxide, Phosphide, and Boride

successfully synthesized stoichiometric δ-and γ-MoN with large crystallite sizes (Wang et al. 2015a). Recently this high-P methodology is extended to the Mo-N system with a focus on nitrogen-rich phases, leading to the discovery of a novel nitride, MoN_2. This novel 3R-MoN_2 adopts a peculiar $4d^{2.5}$ electron configuration with a non-integral valence of +3.5 for Mo ions, and exhibits superior catalytic activities and high hydrogenation selectivity in the hydrodesulfurization of dibenzothiophene compared to MoS_2 (Figure 5). Successful high-P synthesis of nitrides with higher oxidation states demonstrates that high pressure can effectively promote the role of d-electrons in chemical bonding with nitrogen.

Figure 5. HRTEM images and electron diffraction pattern for 3R-MoN_2. And Catalytic activities of 3R–MoN_2 in the hydrodesulfurization (HDS) of dibenzothiophene (DBT). (a) Plate-shaped crystals with laminate edges. (b) Atomic-scale images of the N–Mo–N interlayers, that is, the (003) plane, and (c) the (101) plane. The inset in both (b) and (c) are the enlarged portions of the regions highlighted in yellow. Dotted double red lines indicate the d-spacing for the fingerprints of (003) and (101) planes. Dotted cyan lines in (c) denote the grain boundaries. (d) Selected area electron diffraction, SAED. (e) Temperature-programmed desorption (TPD) of CO to determine the amounts of active Mo sites. Before experiment, the catalysts were pre-exposed to CO gas at 323 K for 60 min to absorb CO molecules and form Mo–CO bonds with active Mo sites, including coordinately unsaturated (1-010) and (101-0) Mo sites. (f) HDS of DBT at 623 K and 6 MPa in H2 atmosphere. (g) Relative DBT concentration, c/c_0, as a function of reaction time t. The normalized pseudo-first-order rate constant, k, was deduced by linear fits. (h) The measured selectivity of direct desulfurization (DDS) and hydrogenation (HYD) in the HDS of DBT.
Source: Reprinted with permission from Wang et al. (2015b). Copyright of American Chemical Society.

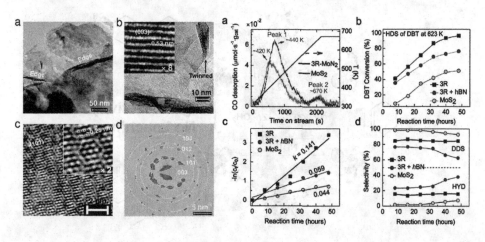

3. THE SYNTHESIS AND CHARACTERIZATION OF Mo(W) OXIDE BASED CATALYSTS

In petroleum and chemical industry, a large number of important chemical reactions are catalyzed by Mo(W) based oxide, such as isomerization, ammoxidation of propene to acrylonitrile, olefin epoxidation, and olefin metathesis reactions. Recently, W(Mo)O$_3$ based catalysts are used to produce the hydrogen via the photo and electric processes. Many novel synthetic methodologies have been employed for making better catalysts. In this section, we only present some examples to depict a profile of the newly progress in the synthesis of Mo(W) oxide based catalysts. More contents are provided in the application part afterward.

Noh et al. (2016) deposited molybdenum (VI) oxide on the Zr-6 node of the mesoporous metal–organic framework NU-1000. Exposure to oxygen leads to a monodisperse, porous heterogeneous Mo-SIM catalyst. It achieved near-quantitative yields of cyclohexene oxide and the ring-opened 1,2-cyclohexanediol for the epoxidation of cyclohexene, and demonstrated no loss in the metal loading before and after catalysis reaction. In contrast, Mo-ZrO$_2$ catalyst led to significant leaching and close to 80 wt % loss of the active species. The stability of Mo-SIM was further confirmed by density functional theory calculations. It is shown that the dissociation of the molybdenum(VI) species from the node of NU-1000 is endergonic (Figure 6).

Figure 6. Metal-organic framework supported molybdenum (VI) oxide catalyst for cyclohexene epoxidation.
Source: Reprinted with permission from Noh et al. (2016). Copyright of American Chemical Society.

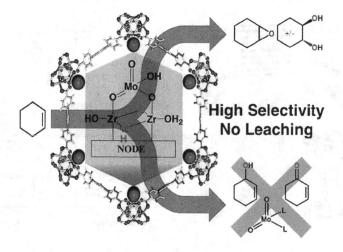

Amakawa et al. (2013) prepared highly dispersed molybdenum oxide on mesoporous silica SBA-15 by anion exchange, resulting in a series of catalysts with different Mo densities (0.2–2.5 Mo atoms nm^2). X-ray absorption, UV/Vis, Raman, and IR spectroscopy indicated that the major surface species are the doubly anchored tetrahedral dioxo MoO_4 units. A steep increase in the catalytic activity was observed in metathesis of propene and oxidative dehydrogenation of propane at 8% of Mo loading, which is attributed to strained configuration of Mo oxide surface species based on DFT calculations and NEXAFS spectra at the O-K-edge.

Kitano et al. (2013) investigated MoO_3/Al_2O_3 acidic catalysts with MoO_3 loadings of 5-30 wt% calcined at high temperatures. It is evidenced that Brønsted acid sites, where acid-catalyzed reactions take place, are generated by calcination at high temperatures. 11 wt% MoO_3/Al_2O_3 calcined at 1073 K exhibited the highest activity owing to the largest numbers of Brønsted acid sites. XPS and Mo K-edge XAFS revealed that molybdenum oxide monolayer domains were stabilized on alumina below 11 wt% of MoO_3 loading. Brønsted acid sites are suggested at boundaries between molybdenum oxide monolayer domains and/or small MoO_3 clusters. When the MoO_3 loading was larger than 11 wt%, some of the Brønsted acid sites on MoO_3/Al_2O_3 was covered with $Al_2Mo_3O_{12}$, resulting in a lowering of the catalytic activity.

Debecker et al. (2012) synthesized MoO_3-SiO_2-Al_2O_3 catalysts consisted of perfectly spherical particles with variable diameter by aerosol processing coupled with surfactant-templated sol–gel method. Spherical particles are produced after the evaporation of water and ethanol contained in spherical aerosol droplets. This is correlated to the templating effect, silica condensation occurs around micelles of surfactant. All particles have the same regularly organized porosity of 1.8-2.0 nm. The catalysts show a high specific surface area and outstanding olefin metathesis activity.

Brookes et al. (2013) synthesized a series of MoO_x-modified Fe_2O_3 catalysts with core-shell structures. It is shown that the octahedral MoO_x layer can stay at the surface even after calcination to 600 °C, which is active and selective for the methanol to formaldehyde synthesis. On the other hand, for the catalyst of MoO_3 nanoparticles deposited on Fe_2O_3, the surficial MoO_3 can react with under-layer Fe_2O_3 to form ferric molybdate $Fe_2(MoO_4)_3$ above 400 °C, which is inert for the selective oxidation.

He et al. (2015) designed the mesoscale structure of molybdenum-vanadium based complex oxide (i.e., Mo-V-M-O, M = Ta, Te, Sb, Nb, etc.) for improve the properties of selective oxidation (Figure 7). They successfully achieved the epitaxial intergrowth between the M1 and M2 phases in this system that are catalytically critical. It was demonstrated that the resulting

Figure 7. Polyhedral structural models and the representative HAADF images of (a) the M1 and (b) the M2 phases in Mo-V-Te-Ta oxide catalysts, viewed at the ab-plane. In the HAADF images, only cation columns are apparent as oxygen has a lower atomic number and is not visible due to dynamic range issues. Black frames in the models represent the unit cells in each structure M_6O_{21} type (M = Ta, Mo or V) units in the M1 phase are highlighted as follows: MO_7 pentagonal bipyramids are highlighted in dark blue and surrounded by edge sharing light blue MO_6 octahedra; gray squares represent the rest of the connecting MO_6 octahedra. The polyhedral network has resulted in heptagonal (M1 phase only, shown in yellow) and hexagonal channels (M1 and M2 phases, highlighted in red). The visibility of Te oxide units inside these channels depend on their occupancies. Te oxide units in the channels are omitted in the structural model in order to make the framework structure clearer. Scheme show the observation of model catalyst then design of optimized structures
Source: Adapted with permission from He et al. (2015). Copyright of American Chemical Society.

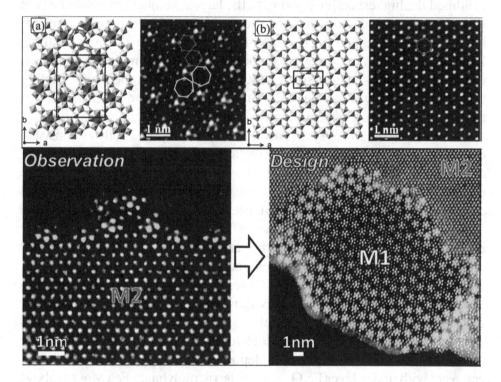

catalyst has improved selectivity for propane ammoxidation owing to the reasonable synergetic interaction.

Phuruangrat et al. (2010) synthesized uniform hexagonal WO_3 (hex-WO_3) nanowires with a diameter of 5–10 nm and lengths of up to several micrometres by a microwave-assisted hydrothermal method. The process was carried out at 150 °C for 3 h in a solution containing $(NH_4)_2SO_4$ as a capping reagent and Na_2WO_4 as the starting material. The aspect ratio and specific surface area of obtained nanowires were 625 and 139 m^2 g^{-1}, respectively. This hex-WO_3 was demonstrated a promising electrocatalyst for the hydrogen evolution reaction (HER) from water.

Guo et al. 2012 deposited nano tungsten oxide (WO_3) particles on the surface of graphene (GR) sheets by a simple sonochemical method. The obtained composite, WO_3@GR, was evidenced the formation of chemical bonds between the nano WO_3 particles and the GR sheets. The average particle size of the WO_3 is around 12 nm on the GR sheets. When used as photo catalyst for water splitting, The WO_3@GR composite with 40 wt% GR inside was twice and 1.8 times as much active as pure WO_3 and mixed-WO_3/GR, respectively, which is due to the synergistic effects of the combined nano WO_3 particles and GR sheets in composite. The chemical bonding between WO_3 and GR minimizes the interface defects, reduces the recombination of the photo-generated electron–hole pairs and enhance the visible light absorption; furthermore, the GR sheets in the WO_3@GR composite enhance electrons transport. These lead to improved photo conversion efficiency.

Herrera et al. (2006) prepared tungsten oxide supported on SBA-15 mesoporous silica through atomic layer deposition (ALD). High dispersions are achieved even at a WO_3 loading of 30 wt%, and enhanced thermal stability is observed, which is attributed mainly to improved resistance toward sintering due to a better interaction of WO_x with the support surface. This is especially beneficial for reactions requiring severe conditions, such as methanol dehydration.

Jiao et al. (2011) grown tungsten trioxide hydrate ($3WO_3 \cdot H_2O$) films directly on fluorine doped tin oxide (FTO) substrate via a facile crystal-seed-assisted hydrothermal method. Scanning electron microscopy (SEM) analysis showed that $3WO_3 \cdot H_2O$ thin films with platelike, wedgelike, or sheetlike nanostructures could be selectively synthesized by adding Na_2SO_4, $(NH_4)_2SO_4$, and CH_3COONH_4 as capping agents, respectively. The film grown using CH_3COONH_4 as a capping agent showed the best solar water splitting activity generating 0.5 mA/cm^2 with the photoconversion efficiency of about 0.3% under simulated solar illumination.

Tungstated zirconia is a robust solid acid catalyst for light alkane (C4–C8) isomerization. However, catalytically active sites remains controversial because of the absence of direct structural imaging information on the various supported WO_x species. Subnanometre Zr–WOx clusters recently are identified as active sites with high-angle annular dark-field imaging of WO_3/ZrO_2 catalysts in an aberration-corrected analytical electron microscope.

This information was used in the design of new catalysts; the activity of a poor catalyst was increased by two orders of magnitude using a synthesis procedure that deliberately increases the number of relevant active species (Zhou et al. 2009).

4. THE SYNTHESIS AND CHARACTERIZATION OF Mo(W) PHOSPHIDE AND BORIDE

Reduction of Mo(W) metal compounds together with phosphate can form metal phosphides under high temperature. Using phosphite, hypophosphite, or phosphine, as well as the plasma reduction of phosphate, lower temperatures are needed and obtained metal phosphide particles are smaller and more active. Organometallic routes allow synthesis of metal phosphide with nano-size (Prins et al. 2012). Like carbide or nitride, metal phosphide synthesized by reduction of precursors in H_2 is usually passivated by passing a dilute oxygen flow (e.g. 1 mol% O_2/He) prior to expose to air. A metal oxide surface layer is formed during passivation, which protects the underlying metal atoms from further oxidation. Before usage, the metal oxide layer can be removed by contacting with H_2.

Xiao et al. (2014) compared Mo, Mo_3P and MoP for HER and evidenced that phosphorization can lead to distinct activities and stabilities in both acid and alkaline media even in bulk form. DFT calculation shows that phosphorization of molybdenum to form MoP introduces an 'H delivery' system which attains nearly zero binding to H at a certain H coverage.

McEnaney et al. (2014) prepared MoP nanoparticles of approximately 4 nm via the solution-phase synthesis technique using $Mo(CO)_6$ and trioctylphosphine as precursors. This nano catalyst are found among the most active known molybdenum-based HER systems with acid stability, as shown in Figure 8.

Yan et al. (2016) anchored cluster-like molybdenum phosphide particles on reduced graphene oxide (MoP/rGO) with high uniformity using phosphomolybdic acid as a molybdenum precursor. The MoP/rGO hybrid exhibits superior electrocatalytic activity towards the hydrogen evolution reaction both in acidic and alkaline media due to the small size and even distribution of the active particles.

Using the open porosity derived from pyrolysis of metal–organic frameworks (MOFs), Yang et al. (2016b) recently deposited the highly dispersive MoO_2 small nanoparticles in porous carbon by chemical vapor deposition (CVD) and HF leaching. Undergoing different treatments, the Mo_2C-, MoN-,

Figure 8. The HRTEM images of MoP particles and the performance of HER ()
Source: Reprinted with permission from McEnaney et al. (2014). Copyright of American Chemical Society.

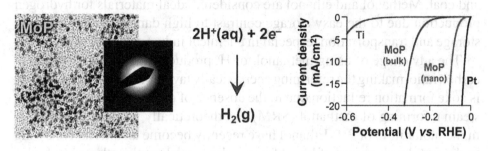

and MoP-decorated carbo-catalysts can be separately prepared. The comparative experiments indicated that the MoP@Porous carbon (MoP@PC) composites exhibited remarkable catalytic activity for HER in 0.5m H_2SO_4 aqueous solution versus MoO_2@PC, Mo_2C@PC, and MoN@PC. These suggests opportunity to synthesize new Mo(W) based MOFs composites for catalytic applications.

Wang et al. (2009) prepared Ni–Mo–B amorphous bimetallic catalysts by chemical reduction of nickel nitrate, ammonium heptamolybdate and sodium borohydride aqueous solution with ultrasonic-assistance. The catalyst presented high catalytic activity of phenol hydrodeoxygenation (HDO). Vrubel and Hu (2012) reported that molybdenum boride (MoB) can be an active HER catalyst.

5. APPLICATION OF Mo(W)-BASED 3D CATALYSTS IN PETROLEUM AND CHEMICAL INDUSTRY

5.1 Hydrogen Production

Hydrogen is one of the most important feed in refining and chemical industry used in upgrade or hydrotreatment of oil streams, hydrogenation and hydrogenolysis processes. With the more stringent specification to transport fuels on the contents of sulfur, olefins and benzene, the H_2 consumption has increased apparently recent years.

Conventionally, the reforming naphtha to produce aromatics provides most of hydrogen needed in petroleum refining. However, with the apparent increase of hydrogen consumption, hydrogen must be acquired via other routes. Several processes such as steam reforming, autothermal reforming, partial reforming, and water gas shift have been applied to extract hydrogen from hydrocarbon compounds such as naphtha, methane, ethanol, methanol and coal. Methanol and ethanol are considered ideal materials for hydrogen production due to the easy storage contrast to high danger of the hydrogen storage and transportation, especial in chemical industry (Jones et al. 2008).

The advantage of using methanol for H_2 production is its high hydrogen/carbon ratio, making the reforming energetically favorable. Another advantage is coke formation reduction due to the absence of carbon-carbon bonds. The steam reforming of methanol (SRM) can theoretically produce H_2 and CO_2 in the molar ratio of 3:1. Ethanol have recently become an alternative owing to the fast development of bio-refinery which produce the ethanol as fuel.

Széchenyi and Solymosi (2007) prepared Mo_2C by the carburization of MoO_3 with a C_2H_6/H_2 mixture, and found that it is an effective catalyst for the decomposition of ethanol and methanol with a high thermal stability. Koós et al. (2008) investigated the promoting effect of potassium on the Mo_2C/C Norit catalyst for the decomposition and reform of methanol, and observed that adding K to Mo_2C/Norit markedly promoted the methanol reforming effectiveness by enhancing the WGS reaction, and the K-modified carbide catalyst had longer-term stability in the SRM. Ma et al (2014a) doped Pt on the carbide surface by carburization of Pt-doped MoO_3 at 700 °C. The obtained catalysts had very high catalytic activity and stability compared with pure β-Mo_2C and iron group metal modified molybdenum carbides for SRM. A methanol conversion of 100% was achieved at a temperature as low as 200 °C and good stability was observed. Doping Ni in molybdenum carbide increased the resistance to the oxidation and carbon deposition, leading to high catalytic activity and stability (Ma et al. 2014a and 2014b). Chen et al. (2016) prepared metal modified molybdenum carbides (M–Mo_2C) (M: Ni, Fe, Co) as the catalysts for SRM, where aqueous solutions of $(NH_4)_6Mo_7O_{24} \cdot 4H_2O$ were mixed with $Ni(NO_3)_2 \cdot 6H_2O$, $Fe(NO_3)_3 \cdot 9H_2O$, and $Co(NO_3)_2 \cdot 6H_2O$, respectively, and then dried to obtain the M–MoO_3. Carburization was carried out in 20% CH_4/H_2 with a temperature-programmed process, followed by the passivation in 1% O_2/Ar. All metal-doped catalysts showed the characteristic of the β phase of the molybdenum carbide and good catalytic activity. Park et al. 2013 prepared the Pt–MoO_x/TiO_2 catalysts through co-impregnation of the precursors and demonstrated unprecedentedly high catalytic activities with extremely low CO selectivities compared with previously reported Pt-based catalysts.

Steam reforming of dimethyl ether (DME) is one of alternatives to produce hydrogen at low temperatures (200–300 °C), which is composed of two consecutive reactions. The first step is the hydration of DME to form methanol over solid acid catalysts: $CH_3OCH_3 + H_2O \rightarrow 2CH_3OH$, and the next step is the steam reforming of methanol $CH_3OH + H_2O \rightarrow 3H_2 + CO_2$. Although the equilibrium conversion of hydration of DME is low (about 20% at 275 °C), high DME conversion may be achievable if methanol formed in the first step is rapidly converted into H_2 and CO_2 by methanol steam reforming catalysts. For conversion of DME into methanol at low temperatures, a strong acid catalyst is preferred but strong acid tends to cause catalyst deactivation by coke formation at the same time. Thus, a compromise is usually needed. Nishiguichi et al. (2006) reported that catalyst with 80 wt% CuO/CeO_2 and 10 wt% WO_3/ZrO_2 can exhibit extended stability. Meanwhile the partly deactivated catalyst can be regenerated by the calcination in air.

Water electrolysis is a well-established method to produce high purity hydrogen. This provides a promising solution to sustainable hydrogen generation (Morales-Guio et al. 2014 and Chen et al. 2013a). Currently, platinum remains the most efficient hydrogen evolution reaction (HER) catalyst in both alkaline and acidic electrolytes, exhibiting both very low overpotential and very good stability. However, due to the low abundance and high expense of Pt, developing efficient noble-metal-free catalysts for HER is critical for supporting the hydrogen production from water electrolysis. Molybdenum-based materials with a diverse range of compositions and structures are among the most heavily studied non-Pt HER catalysts.

Liao et al. (2014) demonstrated the prominent HER performance of the composite of nanoporous Mo_2C nanowires and Mo_2C nanocrystallites. Lin et al. (2016b) developed effective cobalt doping over Mo_2C nanowires to optimize their HER. With Co/Mo ratio of 0.020, the doped catalyst exhibited high activity and good stability in both acidic and basic electrolytes.

Setthapun et al. (2008) reported that Ni-Mo-N catalyst is highly active for HER in alkaline solutions, but is generally unstable in acidic. Chen et al. (2015) reported the acid stable Ni–Mo nitride nanosheets obtained by high temperature ammonia treatment; however, this nitride catalyst only shows moderate HER activity with onset potential of 157mV. McKone et al. (2013) prepared a Ni-Mo nano catalyst by hydrogen reduction of ammonium nickel molybdate, which exhibited very low over-potential of 70 mV in 2M KOH and 80 mV in 0.5M H_2SO_4 at 20 mA/cm^2. However, this catalyst also degraded rapidly in acid. Recently Wang et al. (2016) synthesized a biphasic Ni–Mo–N HER catalyst composed of homogeneously distributed nanocrystals of metallic Ni and $NiMo_4N_5$ which shows very low overpotential in both acid ($\eta 20=53$ mV) and alkaline ($\eta 20=43$ mV) electrolytes and keeps excellent stability. Ternary molybdenum compounds $Co_{0.6}Mo_{1.4}N_2$ are also active and acid-stable for HER (Cao et al. 2013). McCrory et al. (2015) suggested that the Ni-Mo catalyst is one of the best noble metal free HER catalysts in both alkaline and acidic electrolytes.

Amorphous MoP catalysts are proven to catalyze the HER in acidic solutions with overpotentials as low as -90 and −105 mV at current densities of -10 and −20 mA cm^{-2}, respectively (McEnaney et al. 2014). Crystalline MoP catalyst with different nanostructured morphologies also showed excellent HER activities (Xiao et al. 2015 and Xing et al. 2014). They compete with the reported best performances of the nanostructured MoS_2, MoS_x, and Mo_2C HER electrocatalysts.

Photoelectrochemical (PEC) water splitting to produce hydrogen and oxygen has attracted considerable attention owing to its green and sustainable property. Various types of semiconductors, such as titanium dioxide (TiO_2), haematite (α-Fe_2O_3) and tungsten trioxide (WO_3), have been studied for this process (Cristino et al. 2011). Among them, WO_3 and their composites attracts special interests.

Wang et al. (2012) reported that photostability and photoactivity of WO_3 for water splitting can be simultaneously enhanced by introduction of oxygen vacancies into WO_3 in hydrogen atmosphere at elevated temperatures. This is attributed to the formation of substoichiometric WO_{3-x} by hydrogen treatment, which is resistive to the re-oxidation and peroxo-species induced dissolution.

Compared with single semiconductor electrodes, a heterojunction electrode, which contains two or more dissimilar semiconductors can offer more advantages for PEC water splitting. Among the wide variety of heterojunction systems, tungsten trioxide/bismuth vanadate (WO_3/$BiVO_4$) has been one of the most studied (Hong et al. 2011). The combined properties of WO_3 and $BiVO_4$ allow this heterojunction system to have a wider range of wavelengths of photon absorption due to the relatively narrow band gap of $BiVO_4$ and better charge transfer owing to WO_3. Shi et al. (2014) reported the highest photocurrent density to date at 1.23 V versus the reversible hydrogen electrode by using of the bismuth vanadate-decorated tungsten trioxide helical nanostructures due to the combination of effective light scattering, improved charge separation and transportation and an enlarged contact surface area with electrolytes. Liu et al. 2011 fabricated heteronanostructured photoelectrodes by atomic layer deposition of WO_3 and stabilized with a Mn-based oxygen-evolving catalyst. The resulting electrode absorbs photons to create electrons and holes, the separation of which is assisted by the built-in field within the hetero-semiconductor. Electrons are collected, and holes transferred to the catalyst to split H_2O into oxygen and hydrogen.

5.2 Hydrogenation

Numerous studies have shown that carbides of molybdenum and tungsten exhibit catalytic activities for hydrogen reactions, such as CO and CO_2 hydrogenation, hydrogenation of aromatics, ammonia synthesis and methanation (Sabinis et al. 2015b). The turnover rates for such reactions over carbide catalysts can be equal to or greater than noble metal such as Pt, Pd and Ru supported on oxide, and these materials were deemed cheaper alternatives to the noble metal catalysts.

It is known that syngas is converted to light hydrocarbons over molybdenum carbide. The addition of alkalis is of crucial in shifting the products from hydrocarbons to alcohols (Shou et al. 2012). These promoters affect the electron state of Mo_2C. Liakakou and Heracleous (2016) doped K/Mo_2C catalysts with Ni which improved the hydrogenation of CO to higher alcohols. However, considerable deactivation was observed due to the segregation of the mixed Ni-Mo carbide phase and the formation of hydrated oxidic nickel species together with the $K_2Mo_2O_7$ phase.

Previously reported results indicate that the basicity of the support are critical to the production of alcohols and other oxygenated compounds for Mo based catalysts, And acidic sites on supports induce formation of dimethyl ether (DME) and hydrocarbon, thus causing a reduction in alcohol selectivity. However, Acidity is clearly related with increased CO activation (Liakakou et al. 2015). Therefore optimum of higher alcohol synthesis should need a balanced amount of acid sites in catalysts with a trade-off between high conversions and high higher alcohol synthesis (HAS) formation.

Ethanol is primarily produced on yeast fermentation processes using biomass, which is deemed renewable but more expensive. Indirect synthesis of ethanol from syngas is considered as an alternative. This technology includes syngas production, CO oxidative coupling to dimethyl oxalate (DMO), and subsequent hydrogenation to ethanol. Liu et al. (2016) reported that Mo_2C/SiO_2 catalyst is highly active, selective and stable for the hydrogenation of dimethyl oxalate to ethanol at low temperatures of 473 K. Interestingly, the formation of ethanol over the Mo_2C catalyst performs via the novel intermediate methyl acetate instead of ethylene glycol forming over the conventional Cu catalyst, as shown in Figure 9.

1,3-propanediol (1,3-PD) is a valuable chemical used in the synthesis of polytrimethylene terephthalates (PTT) and in the manufacture of polyurethanes and cyclic compounds. 1,3-PD is currently produced from petroleum derivatives such as ethylene oxide (Shell route) or acrolein (Degussa-DuPont route) by chemical catalytic routes. Much attention has been recently focused

Figure 9. The reaction pathway of the ethanol synthesis form syngas

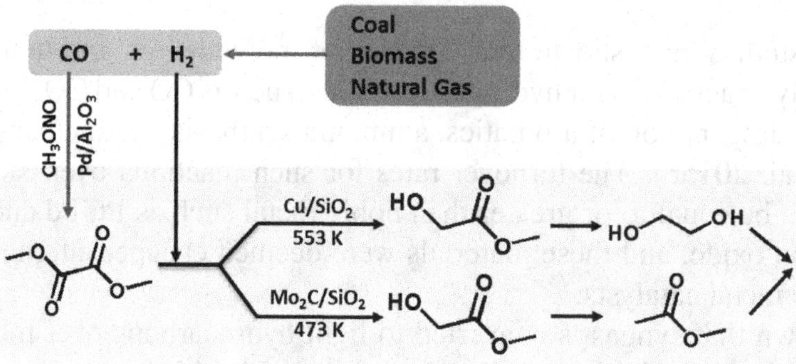

The coupling of CO with methanol to form dimethyl oxalate

$$2CO + 2CH_3ONO \rightarrow CH_3OOCCOOCH_3 + 2NO$$

$$2NO + 2CH_3OH + O_2 \rightarrow 2CH_3ONO + H_2O$$

$$2CO + 2CH_3OH + O_2 \rightarrow CH_3OOCCOOCH_3 + H_2O$$

on the development of effective methods for the production of 1,3-PD from glycerol owing to glycerol is largely produced in the biodiesel industry; approximately 1 ton of a crude glycerol is formed for every 10 tons of biodiesel produced. Glycerol hydrogenolysis catalyzed by $Pt/WO_3/ZrO_2$ afforded 1,3-PD in the yields up to 24%. The catalytic activities and the selectivity toward 1,3-PD were remarkably affected by the type of support, loaded noble metal and preparation procedure (Kurosaka et al. 2008).

Functionalized arylamines are industrially important organic intermediates for pharmaceuticals, dyestuffs, functional polymers etc. They are now mainly produced by selective hydrogenation of the corresponding nitroarenes. The chemoselective synthesis of functionalized arylamines over Co modified metal carbides can be comparable to those of precious metals, which is ascribed to the remarkably synergistic effect between Co and Mo_2C (Zhao et al. 2014).

CO_2 reduction to fuel or chemicals decreases the emission of greenhouse gas and facilitates the cycle utilization of CO_2. Chen et al. (2012) synthesized an ultrathin, single-crystal WO_3 nanosheet with ~4–5 nm in thickness using a solid–liquid phase arc discharge route in an aqueous solution. Size-quantization effects in this ultrathin nanostructure alter the WO_3 band gap, leading to enhanced performance for photocatalytic reduction of CO_2 to CH_4 in the presence of water that do not exist in its bulk form. Xi et al. (2012) prepared oxygen-vacancy-rich ultrathin $W_{18}O_{49}$ nanowires up to several micrometers long by a facile one-pot synthesis. The nanowires show an unexpected ability to reduce carbon dioxide to methane photochemically due to its large quantities of oxygen vacancies.

Xie et al. (2012) reported a facile route of synthesizing a quasi-cubic-like WO_3 crystal with a nearly equal percentage of {002}, {200} and {020} facets, and a rectangular sheet-like WO_3 crystal with predominant {002} facet by controlling acidic hydrolysis of crystalline. It is found that the quasi-cubic-like WO_3 crystal with a deeper valence band maximum shows a much higher O_2 evolution rate in photocatalytic water oxidation than the rectangular sheet-like WO_3 crystal. On the other hand, the latter with an elevated conduction band minimum of 0.3 eV is able to photoreduce CO_2 to generate CH_4 in the presence of H_2O vapor.

5.3 Selective Oxidation

Ethylene is the most important building blocks in the chemical industry, ranking first with respect to volume, which is used to synthesize polymers, styrene, ethylene oxide, vinyl chloride and vinyl acetate monomers, functionalized hydrocarbons (i.e., dichloroethane, ethylbenzene, acetaldehyde, and ethanol), and many other basic and intermediate chemical products.

High-temperature pyrolysis in the presence of diluting steam is the most established industrial process for the manufacture of ethylene now. The availability of biomass and ethane in shale gas and refinery gas has advanced the interest in alternative processes for ethylene production, including the dehydration of ethanol (thus enabling the utilization of biomass-derived feeds) and the oxidative dehydrogenation (ODH) of ethane.

Mo oxides are active for the ODH of ethane. However, mixed oxide based systems may be better suited than single oxides (Botella et al. 2006). It was suggested that Mo/V mixed-oxide system with an orthorhombic Mo_3VO_x structure is very active for ethane ODH. The high reactivity has been associated to the formation of pentagonal Mo_6O_{21} units, which form micropores constituted by heptagonal channels. The catalytic reaction is speculated proceeding within these pores (Gaertner et al. 2013). Mo/V-based systems can be promoted with addition of Al, Ga, Bi and Ce (Chu et al. 2015). More-complex systems, such as $MoVNbO_x$ and $MoVSbTeO_x$, are reported very active and selective in the ODH of ethane (Cheng et al. 2015).

Among all the catalytic systems above, the mixed metal oxide catalyst, $MoVNbTeO_x$ is one of the most outstanding catalysts for ODH of ethane. This catalyst usually consists of M1 and M2 crystalline phases and minor amounts of other phases such as Mo_5O_{14}-type structures or binary MoV and

MoTe oxides. M1 is an orthorhombic phase with space Pba2 group, and M2 is a pseudo-hexagonal phase with P6mm space group. Studies have revealed that the M1 phase possesses the only V^{5+} ions as ethane activation sites, which dominates the performance of MoVNbTeOx catalysts (Melzer et al. 2013). Chu et al. (2015) successfully prepared pure-phase M1/CeO_2 nanocomposite catalysts by physical mixing or sol–gel method. The ODH experimental results show that the introduction of CeO_2 can increase the abundance of V^{5+} on the catalyst surface by a self-redox solid-state reaction during activation at 400 °C in air. The nanocomposite catalyst consisting of M1 particles and 4.4 nm CeO_2 exhibits the best productivity of 0.66 kgC_2H_4/kgcat h at 400 °C with 20% lower cost than pure-phase M1 catalysts. Until now, however, the ethylene productivity with the reported catalyst cannot meet the industrial requirements yet for commercialization (1.00 kgC_2H_4/kgcat h).

Molybdenum oxide based catalysts are frequently used in selective oxidation or ammoxidation of light alkanes and alkenes. Although the most simple molybdenum oxide, orthorhombic α-MoO_3, exhibits poor catalytic activity, addition of other transient metals like tungsten or vanadium leads to an increase in activity and selectivity towards objective products.

Mo-V oxides presents the activity for the (ammo) oxidation of alkanes (Lin et al. 2001). Lopez-Medina et al. (2011) reported preparation of nanoscaled supported MoVNbTeO catalysts for the propane ammoxidation. The best-supported catalyst afforded ca. 50% acrylonitrile yield with 80% propane conversion at 450 °C. Moreover, the activity per gram of MoVNbTeO catalyst increases fourfold upon stabilization of nanoparticles compared with the bulky ones.

Holmberg et al. (2006) studied substitutions in the M2 phase of the Mo-V-Nb-Te-oxide system. In the pseudo-hexagonal M2 structure of (TeO)$Mo_2V_1O_9$, They can substitute the whole system by Mo without the basic structure change. However, Ti can replace merely half of the V, and Ce replace about 30% of the Te. Moreover, substitution of Fe for V and Nb for both Mo and V is possible up to 20–30%. The activity data for propene ammoxidation reveal that substitution in the M2 structure of W for Mo and Ce for Te both give higher specific activity and improved selectivity to acrylonitrile at the expense of acrolein. Replacement of V by Ti or Nb exerts no significant influence on the product distribution, but increases the activity. Fe substitute for V seems to lead acrolein burning. With regard to both activity and selectivity to acrylonitrile formation, a catalyst with 70% of the Mo replaced by W exhibits the best performance.

Shiju et al. (2007) reported that crystalline Mo-V-Te-Nb-O oxide with orthorhombic M1 phase structure can be synthesized rapidly and reproducibly by microwave heating. The obtained material exhibits very high catalytic activity for direct ammoxidation of propane to acrylonitrile.

Acrylic acid (AA) and its esters are important bulk chemicals with a broad application in the polymer industry, for example as super-absorbers, fabrics, adhesives and paints. The global annual production of crude AA is about 4 million tons, mostly based on the process of two-step propene oxidation. In the first step, propene is selectively oxidized with air on a Bi/Mo-catalyst to acrolein (ACR). And water is formed as a coproduct. Subsequently, ACR, water and additional air are fed into the second step without further purification. It was reported that the rate of the acrylic acid formation can significantly enhance in the presence of water (Jekewitz et al. 2012). The second step is achieved over a modified Mo/V/W-mixed oxide catalysts. Currently, a selectivity of over 90% can be achieved with industrial optimized catalyst systems. Carbon oxides, such as CO and CO_2, are produced as by-products from the combustion of ACR and AA (Figure 10). The main components Mo, V, W of these catalysts have complicated effects. MoO_3 has proved to be inactive, V_2O_5 is active but unselective and WO_3 is inert. Bi-metallic V and Mo oxides also show activity and selectivity of propene oxidation. Structure types Mo_4VO_{14} and Mo_3VO_{11} are deduced as active phases (Kunert et al. 2004). Amorphous and oxygen-deficient defects are known to be essential for selective oxidations (Giebeler et al. 2010). During the second step, acrolein exchanges its oxygen with oxygen in the mixed oxide catalyst that is evidenced by the kinetics experiment of the isotopes (Drochner et al. 2014).

Outstanding catalytic activity for the selective oxidation of acrolein (90.2% yield at 463 K) is observed over a crystalline metal oxide, Mo_3VO_x ($x \leq 11.1$). The catalyst is synthesized from a solution containing pentagonal units of $Mo(Mo_5O_{27})$, which further react with molybdenum and vanadium species to form a 3D metal oxide catalyst (Sadakane et al. 2007).

Figure 10. Schematic the oxidation of acrolein on Mo(W) based oxides

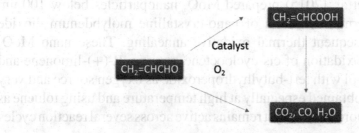

Mo based oxide can also directly catalyze the oxidation of unsaturated molecules to carboxylic acids (Solsona et al. 2004). Commercial process using 1.5% wt.Pd-promoted heteropolyacid (HPA) catalysts for ethene oxidation was reported to give approximately 80% acetic acid selectivity at low temperatures (ca. 420 K). Multicomponent oxides promoted by noble metals ($Mo_1V_{0.396}Nb_{0.128}Pd_{3.848*10^{-4}}$) gave high acetic acid selectivity (ca. 80%) using ethene as a reactant. Their high stability allowed their use as catalysts at relatively high temperatures (560 K), which led to higher acetic acid productivities than on polyoxometalate-based catalysts (Karim and Sheikh, 2000).

Direct selective oxidation of alkane is challenge owing to high energy of C-H bond. Liu et al. 2016 synthesized a series of molybdenum oxide incorporated mesoporous silica catalysts (Mo-KIT-6) by a one-pot co-assembly method for propane selective oxidation to acrolein. The molybdenum substituted into the framework of the KIT-6 support forming highly dispersed active sites that possess appropriate redox properties and a strong ability to resist coke formation. Addition of K to Mo-KIT-6 catalyst further promoted the yield of acrolein to 25.9%.

Palladium-promoted multicomponent metal oxides show unprecedented reactivity in ethane oxidation to acetic acid with the selectivity up to about 90% (Li et al. 2007). Precipitation of Mo-V-Nb active oxides in the presence of colloidal TiO_2 led to much higher active surface areas than in unsupported oxides. Palladium increased the rate of ethane oxidation to acetaldehyde, while active Mo-V-Nb oxides rapidly oxidized the acetaldehyde intermediates to stable acetic acid products, thus preventing acetaldehyde combustion pathway, which is prevalent on monofunctional palladium-based catalysts.

Olefin epoxidation is a major field of research in the preparation of relevant building blocks for organic synthesis. Among other metal systems, Mo-catalyzed olefin epoxidation has received interest from both academic and industrial research laboratories. Since the first example of a molybdenum oxo complex catalyzing the epoxidation of alkenes with peroxides such as organic hydroperoxides and hydrogen peroxide, a variety of systems has been developed. Bento et al. (2015) reported that olefin epoxidation can be achieved over metal oxide nanoparticles MoO_2 with tremella-like morphology. Fernandes et al. (2015) prepared MoO_3 nanoparticles below 100 nm through solvothermal synthesis of nano-crystalline molybdenum dioxide (MoO_2) and subsequent thermal oxidative annealing. These nano MoO_3 were used for epoxidation of cis-cyclooctene, styrene, R-(+)-limonene and trans-hex-2-en-1-ol with tert-butylhydroperoxide as oxygen source and very high yields were obtained especially at high temperature and using toluene as solvent. Furthermore, the catalyst remains active across several reaction cycles with virtually no loss of activity. Pang et al. (2014) reported MoO_3-Bi_2SiO_5/SiO_2 complex catalysts with a Mo/Bi molar ratio of 5 for the epoxidation of propylene by O_2. A propylene oxide selectivity of 55.14% were obtained at 0.15 MPa, 673 K, and flow rates of $C_3H_6/O_2/N_2=1/4/20$ $cm^3 min^{-1}$. NH_3-TPD results indicate that the surface acid sites are necessary for the high catalytic activity. The epoxidation involves an allylic radical generated at the molybdenum oxide species and the activation of O_2 at the bismuth oxide.

It is estimated that formalin production is around 20 Mton a year (when measured as 37 wt% formaldehyde in water). Two main processes are currently used. The older one is through dehydrogenation of methanol over a silver catalyst. And the more recent process involves the oxidative dehydrogenation of methanol over an iron molybdate catalyst. This reaction is of the Mars–van Krevelen type where surface lattice oxygen is the active species (Bowker et al. 2015). Gaseous oxygen is indirectly involved for re-oxidizing the catalyst surface (House et al. 2007).

Caro et al. 2012 developed the selective catalytic oxidation of ethanol to acetaldehyde in air over a catalyst with molybdenum oxide supported on titania which gave high selectivity to acetaldehyde (70%–89%, depending on the Mo loading) at 150 °C and ambient atmosphere. Wang et al. (2009) demonstrated that the Mo-V-O crystalline oxide could catalyze the oxidation of a series of alcohols, with carbonyl compounds as the major products.

Bio-glycerol is an important platform raw material for the production of numerous fine chemicals, pharmaceuticals, cosmetics, plastics, etc. It is now obtained as a by-product in the production of biodiesel. Acrylic acid production from glycerol is deemed promising. However, the reaction protocol for the conversion of glycerol to acrylic acid via an acrolein intermediate always suffers from fast catalyst deactivation and thus cannot be commercialized. Li et al. (2016) suggested a novel two-step protocol for the conversion of glycerol to acrylic acid: glycerol deoxydehydration (DODH) by formic acid to allyl alcohol, followed by oxidation to acrylic acid in the gas phase over Mo–V–W–O catalysts. In the second step, the mesoporous silica-supported Mo–V–W–O catalysts showed superb stability on time stream under the optimal reaction conditions with an overall acrylic acid yield of 80%. This study suggests a potential process for the large-scale production of acrylic acid from biorenewable glycerol (Figure 11).

Mo based catalysts have been investigated for other selective oxidation. Multi-component MoVNbTe mixed oxides, in contrast to three-component MoVNb and MoVTe oxides show exclusive efficiency in selective oxidation of ethanol by molecular oxygen to obtain acetic acid (Sobolev et al. 2011). Zhang et al. (2016a) investigated the MoO_3–SnO_2 catalysts for glycol di-

Figure 11. Conversion of glycerol to acrylic acid via gas phase catalytic oxidation of an allyl alcohol intermediate
Source: Reprinted with permission from Li et al. (2016). Copyright of American Chemical Society.

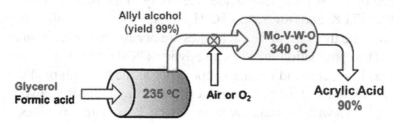

methyl ether (DMET) to 1,2-propanediol (PDO). Three MoO_3/SnO_2 catalysts were prepared respectively with orthorhombic (α), monoclinic (β) and hexagonal (h) MoO_3 crystalline phases. The highest PDO selectivity was always obtained over the h-MoO_3–SnO_2 catalyst and the lowest PDO selectivity was always obtained over the β-MoO_3–SnO_2 catalyst. Zhang et al. (2016b) developed an efficient and economically viable process for the conversion of dimethyl ether (DME) into high value methyl formate (MF); they found that a high degree of MoO_3 and SnO_2 interface contact has a strong positive effect on the conversion of the methoxyl intermediate to MF in the dimethyl ether (DME) oxidation reaction.

5.4 Acid Catalysis

Solid acid catalysts play an important role in a great number of petroleum and chemical processes, such as catalytic cracking, isomerization, alkylation, and hydration and dehydration, due to high activity, convenient separation, and lesser corrosion of reactors and environmental problems (Kumar et al. 2013). Molybdenum or Tungsten oxide-based materials such as MoO_3/Al_2O_3 or ZrO_2 (Kitano et al. 2013), WO_3/Al_2O_3 (Chen et al. 2010), WO_3/TiO_2 (Zhang et al. 2011), WO_3/SiO_2 (Spamer et al. 2003), and WO_3/ZrO_2 (Song et al. 2013) are considered promising solid acid catalysts, owing to their strong acidity and thereby notable activity for various catalytic reactions.

Kim et al. (2007) supported WO_3 on Al_2O_3, Nb_2O_5, TiO_2, and ZrO_2 by impregnation of aqueous ammonium metatungstate, $(NH_4)_{10}W_{12}O_{41} \cdot 5H_2O$. The relative acidity of the different tungsten oxide components are found depending on the specific oxide support. For supported WO_3/Al_2O_3 catalysts, the surface WO_x species are more active than the crystalline WO_3 particles. For the other supported WO_3 catalysts, however, the crystalline WO_3 particles are more active than the surface WO_x species. These reflect the important effect of the oxide support on the acidic activity of the surface WO_x species.

Zirconia-supported W or Mo oxide catalysts have received much attention due to their industrial application for converting C4–C8 paraffins to highly branched species to upgrade the gasoline octane number (Ross-Medgaarden et al. 2008). Barton et al. (1999) studied the roles of Pt and hydrogen on WO_3-ZrO_2 catalyst in the n-heptane isomerization and concluded that Pt/WO_3-ZrO_2 had a higher activity than that of Pt/SO_3-ZrO_2 due to the availability of hydrogen on the surface of WO_3-ZrO_2 to suppress cracking process. Compared with other sulfated zirconia, WO_3/ZrO_2 is more stable under reduction conditions and shows less loss of active species. Surprisingly, the catalytic activity of Pt/MoO_3-ZrO_2 was found inferior to that of MoO_3-ZrO_2 for the C5-C7 linear alkane hydroisomerization, although the Pt/MoO_3-ZrO_2 had higher hydrogen uptake capacity (Triwahyono et al. 2013).

Tagusagawa et al. (2009) found that the protonated, layered transition-metal oxide $HNbMoO_6$ exhibits remarkable catalytic activity for acid-catalyzed reactions by utilizing the strong acid sites present in its interlayers. The acid amount was proportional to the amount of protons in the formula, $H_{1-x}Nb_{1-x}Mo_{1+x}O_6$, and half the protons contributed to effective acid sites. The acid amount and acid strength increased with increasing Nb and Mo concentration, respectively. Layered $H_{1.1}Nb_{1.1}Mo_{0.9}O_6$ functioned as the most active solid acid catalyst for Friedel-Crafts alkylation of toluene. On the other hand, $H_{0.9}Nb0_{.9}Mo_{1.1}O_6$ exhibited the highest activity for hydrolysis of cellobiose and esterification of lactic acid due to the high acid strength. The improvement of catalytic activity for the desired acid-catalyzed reactions can be achieved by adjustment of the metal ratio of layered $HNbMoO_6$, which influenced both the acid amount and the acid strength.

Conversion of glycerol into acrolein and hydroxyacetone (acetol) is considered as an alternative route replacing the oxidation of propylene. The synthesis of acrolein proceeds via a double dehydration reaction. The other product of the dehydration of glycerol is acetol, which is an intermediate for fine chemicals and pharmaceutics also. Chai et al. (2009) reported that 12-tungstophosphoric acid supported on zirconia can almost fully convert glycerol with selectivity of 70–75% to acrolein at 330 °C. Different supported heteropoly acids have been used as catalysts for the gas-phase dehydration of glycerol. Selectivities of 75–96% to acrolein and conversions of 80–95% were achieved at lower temperature between 275 and 285 °C (Tsukuda et al. 2007, Atia et al. 2008, and Alhanash et al. 2010). Now the selective catalytic conversion to acrolein and/or hydroxyacetone according to demand remains very challenging.

5.5 Metathesis Reaction

The current yearly production of C2-C4 olefins exceeds 200 million tons. Propene production by cross-metathesis of ethene and 2-butenes is an economic strategy to satisfy the increasing propene demand. Silica-supported tungsten oxide catalysts are currently employed at high temperature (>573 K). Recent studies suggest that Mo oxides supported on acidic materials (e.g., silica-alumina or SBA-15) are more active in metathesis of propene. Amakawa et al. (2012 and 2015) showed that the active Mo(VI)–alkylidene moieties are generated in situ by surface reaction of molybdenum oxide precursor species with the substrate molecule itself. The active site formation involves sequential steps that require multiple catalytic functions. Protonation of propene to surface Mo(VI)–isopropoxide species occurs on surface Brønsted acid sites; subsequently isopropoxide is oxidized to acetone by the reduction of Mo(VI), leaving naked Mo(IV) sites after desorption of acetone; oxidative addition of another propene molecule yields the active Mo(VI)–alkylidene species. This view is quite different from the older one-step mechanism and has been implemented for catalyst improvement. For example, heat treatment after the initial propene adsorption have doubled the catalytic activity due to accelerating the oxidation, desorption and capturing steps, which demonstrated the merit of knowledge-based strategies in heterogeneous catalysis.

Lwin et al. (2016) investigated WO_x/SiO_2 catalysts for propylene metathesis as a function of tungsten oxide loading and temperature. In situ Raman spectroscopy under dehydrated conditions revealed that below 8% WO_x/SiO_2, only surface WO_x sites are present on the silica support. The activation process produces highly active WO_x site that catalyze olefin metathesis at ~150–250 °C. For 8% WO_x/SiO_2 and higher WO_3 loading, crystalline nanoparticles (NPs) of WO_3, which are not active for propylene metathesis, are identified. The acid character of the surface WO_x sites (Lewis) and WO_3 NPs (Brønsted) is responsible for formation of undesirable reaction products (C4–C6 alkanes and dimerization of C2= to C4= (Figure 12).

5.6 Production and Upgrade of Bio-Diesel

Fats and plant oils, especially present in nonedible or waste feedstocks, are a potential renewable feedstock for transportation diesel fuels. The relatively high oxygen content and acidity (from the free fat acid) of this feedstock compared with fossil fuels, however, lead to several drawbacks as a fuel, such as corrosive properties and higher viscosity. The transesterification of vegetable oils has proven a viable alternative for converting bio available

Figure 12. The schematic reaction processes occurred on different sites in WO_x/SiO_2 catalysts.
Source: Reprinted with permission from Lwin et al. (2016). Copyright of American Chemical Society.

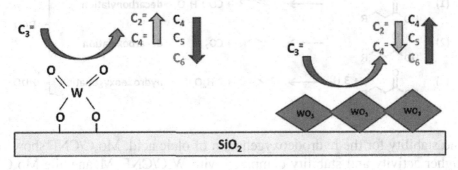

triglycerides into biodiesel fuel. To synthesize fatty acid methyl and ethyl ester, strong homogeneous bases such as KOH are commercially employed with methanol or ethanol as reagent. However, using heterogeneous acid catalysis may provide operation that is more convenient. By the sol–gel process, Bail et al. (2013) synthesized Mo-containing silica catalysts with high concentration of acid sites that achieved near 95% conversion of fatty acids for the esterification of methanol or ethanol. The solid catalysts were highly resistant to leaching and possible to recover and recycle. Zhang et al. (2013) modified Zr–Mo oxide with stearic acid. Under the optimal reaction conditions of methanol/oleic acid molar ratio = 10:1 at 180 °C and 2 h reaction time with 4 wt% of catalyst loading, the conversion of oleic acid reached 94.2%. Mouat et al. (2016) prepared a catalyst with single-site molybdenum dioxo on high-surface-area activated carbon. It catalyzes the transesterification of a variety of esters and triglycerides with ethanol, exhibiting high activity at moderate temperatures (60-90 °C) with negligible deactivation.

DO (deoxygenation) is another promising alternative to upgrade vegetable oils to hydrocarbons in the diesel range and reduce engine compatibility issues. The produced fuels by this process outperform fossil-based ones with cetane numbers ranging from 85 to 99, compared with 45-55 of petroleum diesel. In addition, DO of vegetable oils is an important route to synthesize higher value chemicals by tuning selectivity toward olefins or alcohols (van der Klis et al. 2012).

The deoxygenation mainly occurs via three different pathways: decarbonylation, decarboxylation, and hydrodeoxygenation (Figure 13). W and Mo carbide catalysts are among the most promising catalysts in the deoxygenation of vegetable fats/oils. Hollak et al. (2013) conducted a comparison between carbon nanofiber-supported W_2C and Mo_2C catalysts on activity, selectivity

Figure 13. Pathways of the deoxygenation (DO) of fatty acids
Source: Reprinted with permission from Hollak et al. (2013). Copyright of American Chemical Society.

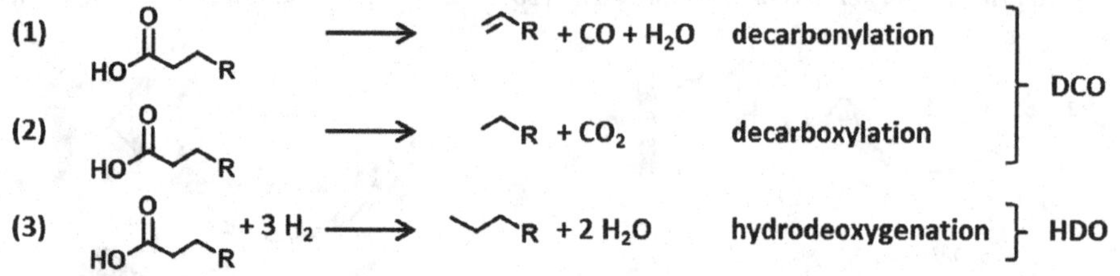

(1) $HOOC-CH_2-R \longrightarrow {\wedge}R + CO + H_2O$ decarbonylation ⎤
(2) $HOOC-CH_2-R \longrightarrow {\wedge}R + CO_2$ decarboxylation ⎦ DCO
(3) $HOOC-CH_2-R + 3H_2 \longrightarrow {\vee\wedge}R + 2H_2O$ hydrodeoxygenation ⎤ HDO

and stability for the hydrodeoxygenation of oleic acid. Mo_2C/CNF showed higher activity and stability compared with W_2C/CNF. Meanwhile Mo_2C/CNF was more selective toward paraffins and W_2C/CNF selective toward olefins. Han et al. (2011) synthesized ordered mesoporous carbon supported molybdenum carbide catalysts by a facile one-pot method. The supported Mo carbide phase could be regulated from Mo_2C to MoC by changing the amount of Mo precursor from less than 2% to more than 5%. MoC catalyst exhibited a high product selectivity and excellent resistance to leaching in the conversion of vegetable oils into diesel-like hydrocarbons compared to Mo_2C catalyst.

Catalysts based on Mo(W) carbide, nitride, phosphide, boride and oxide have been demonstrated promising performances in many different types reactions, including hydrogenation,, selective oxidation, acidic catalysis, and metathesis etc. And they can play important roles in CO_2 conversion, water splitting and biomass conversion which are the main concerns in the sustainable energy and economy. However, many researches are still required for the application of these catalysts for industry processes.

REFERENCES

Alhanash, A., Kozhevnikova, E. F., & Kozhevnikov, I. V. (2010). Gas-phase dehydration of glycerol to acrolein catalysed by caesium heteropoly salt. *Applied Catalysis A, General*, *378*(1), 11–18. doi:10.1016/j.apcata.2010.01.043

Amakawa, K., Kroehnert, J., Wrabetz, S., Frank, B., Hemmann, F., Jaeger, C., & Trunschke, A. et al. (2015). Active Sites in Olefin Metathesis over Supported Molybdena Catalysts. *ChemCatChem*, *7*(24), 4059–4065. doi:10.1002/cctc.201500725

Amakawa, K., Sun, L., Guo, C., Haevecker, M., Kube, P., Wachs, I. E., ... Trunschke, A. (2013). How Strain Affects the Reactivity of Surface Metal Oxide Catalysts. Angewandte Chemie-International Edition, 52(51), 13553–13557.

Amakawa, K., Wrabetz, S., Kroehnert, J., Tzolova-Mueller, G., Schloegl, R., & Trunschke, A. (2012). In Situ Generation of Active Sites in Olefin Metathesis. *Journal of the American Chemical Society, 134*(28), 11462–11473. doi:10.1021/ja3011989 PMID:22703234

Atia, H., Armbruster, U., & Martin, A. (2008). Dehydration of glycerol in gas phase using heteropolyacid catalysts as active compounds. *Journal of Catalysis, 258*(1), 71–82. doi:10.1016/j.jcat.2008.05.027

Bail, A., dos Santos, V. C., de Freitas, M. R., Ramos, L. P., Schreiner, W. H., Ricci, G. P., & Nakagaki, S. et al. (2013). Investigation of a molybdenum-containing silica catalyst synthesized by the sol-gel process in heterogeneous catalytic esterification reactions using methanol and ethanol. *Applied Catalysis B: Environmental, 130*, 314–324. doi:10.1016/j.apcatb.2012.11.009

Barton, D. G., Soled, S. L., Meitzner, G. D., Fuentes, G. A., & Iglesia, E. (1999). Structural and Catalytic Characterization of Solid Acids Based on Zirconia Modified by Tungsten Oxide. *Journal of Catalysis, 181*(1), 57–72. doi:10.1006/jcat.1998.2269

Bento, A., Sanches, A., Medina, E., Nunes, C. D., & Vaz, P. D. (2015). MoO_2 nanoparticles as highly efficient olefin epoxidation catalysts. *Applied Catalysis A, General, 504*, 399–407. doi:10.1016/j.apcata.2015.03.024

Botella, P., Dejoz, A., Lopeznieto, J., Concepcion, P., & Vazquez, M. (2006). Selective oxidative dehydrogenation of ethane over MoVSbO mixed oxide catalysts. *Applied Catalysis A, General, 298*, 16–23. doi:10.1016/j.apcata.2005.09.018

Bowker, M. (2015). Rules for Selective Oxidation Exemplified by Methanol Selective Oxidation on Iron Molybdate Catalysts. *Topics in Catalysis, 58*(10–11), 606–612. doi:10.1007/s11244-015-0399-4

Brookes, C., Wells, P. P., Cibin, G., Dimitratos, N., Jones, W., Morgan, D. J., & Bowker, M. (2014). Molybdenum Oxide on Fe_2O_3 Core-Shell Catalysts: Probing the Nature of the Structural Motifs Responsible for Methanol Oxidation Catalysis. *ACS Catalysis, 4*(1), 243–250. doi:10.1021/cs400683e

Cao, B., Veith, G. M., Neuefeind, J. C., Adzic, R. R., & Khalifah, P. G. (2013). Mixed Close-Packed Cobalt Molybdenum Nitrides as Non-noble Metal Electrocatalysts for the Hydrogen Evolution Reaction. *Journal of the American Chemical Society, 135*(51), 19186–19192. doi:10.1021/ja4081056 PMID:24175858

Caro, C., Thirunavukkarasu, K., Anilkumar, M., Shiju, N. R., & Rothenberg, G. (2012). Selective Autooxidation of Ethanol over Titania-Supported Molybdenum Oxide Catalysts: Structure and Reactivity. *Advanced Synthesis & Catalysis, 354*(7), 1327–1336. doi:10.1002/adsc.201000841 PMID:23396482

Chai, S.-H., Wang, H.-P., Liang, Y., & Xu, B.-Q. (2009). Sustainable production of acrolein: Preparation and characterization of zirconia-supported 12-tungstophosphoric acid catalyst for gas-phase dehydration of glycerol. *Applied Catalysis A, General, 353*(2), 213–222. doi:10.1016/j.apcata.2008.10.040

Chen, W.-F., Iyer, S., Iyer, S., Sasaki, K., Wang, C.-H., Zhu, Y., & Fujita, E. et al. (2013b). Biomass-derived electrocatalytic composites for hydrogen evolution. *Energy & Environmental Science, 6*(6), 1818–1826. doi:10.1039/c3ee40596f

Chen, W.-F., Sasaki, K., Ma, C., Frenkel, A. I., Marinkovic, N., Muckerman, J. T., ... Adzic, R. R. (2012). Hydrogen-Evolution Catalysts Based on Non-Noble Metal Nickel-Molybdenum Nitride Nanosheets. *Angewandte Chemie International Edition, 51*(25), 6131–6135.

Chen, W.-F., Wang, C.-H., Sasaki, K., Marinkovic, N., Xu, W., Muckerman, J. T., & Adzic, R. R. et al. (2013a). Highly active and durable nanostructured molybdenum carbide electrocatalysts for hydrogen production. *Energy & Environmental Science, 6*(3), 943–951. doi:10.1039/c2ee23891h

Chen, X., Clet, G., Thomas, K., & Houalla, M. (2010). Correlation between structure, acidity and catalytic performance of WO_x/Al_2O_3 catalysts. *Journal of Catalysis, 273*(2), 236–244. doi:10.1016/j.jcat.2010.05.018

Chen, X., Zhou, Y., Liu, Q., Li, Z., Liu, J., & Zou, Z. (2012). Ultrathin, Single-Crystal WO_3 Nanosheets by Two-Dimensional Oriented Attachment toward Enhanced Photocatalystic Reduction of CO_2 into Hydrocarbon Fuels under Visible Light. *ACS Applied Materials & Interfaces, 4*(7), 3372–3377. doi:10.1021/am300661s PMID:22738275

Chen, Y., Choi, S., & Thompson, L. T. (2016). Ethyl formate hydrogenolysis over Mo_2C-based catalysts: Towards low temperature CO and CO_2 hydrogenation to methanol. *Catalysis Today, 259*, 285–291. doi:10.1016/j.cattod.2015.08.021

Cheng, M.-J., & Goddard, W. A. III. (2015). In Silico Design of Highly Selective Mo-V-Te-Nb-O Mixed Metal Oxide Catalysts for Ammoxidation and Oxidative Dehydrogenation of Propane and Ethane. *Journal of the American Chemical Society, 137*(41), 13224–13227. doi:10.1021/jacs.5b07073 PMID:26423704

Chu, B., An, H., Nijhuis, T. A., Schouten, J. C., & Cheng, Y. (2015). A self-redox pure-phase M1 $MoVNbTeO_x/CeO_2$ nanocomposite as a highly active catalyst for oxidative dehydrogenation of ethane. *Journal of Catalysis, 329*, 471–478. doi:10.1016/j.jcat.2015.06.009

Cristino, V., Caramori, S., Argazzi, R., Meda, L., Marra, G. L., & Bignozzi, C. A. (2011). Efficient Photoelectrochemical Water Splitting by Anodically Grown WO_3 Electrodes. *Langmuir, 27*(11), 7276–7284. doi:10.1021/la200595x PMID:21542603

Cui, W., Cheng, N., Liu, Q., Ge, C., Asiri, A. M., & Sun, X. (2014). Mo_2C Nanoparticles Decorated Graphitic Carbon Sheets: Biopolymer-Derived Solid-State Synthesis and Application as an Efficient Electrocatalyst for Hydrogen Generation. *ACS Catalysis, 4*(8), 2658–2661. doi:10.1021/cs5005294

Cui, X., Li, H., Guo, L., He, D., Chen, H., & Shi, J. (2008). Synthesis of mesoporous tungsten carbide by an impregnation–compaction route, and its NH_3 decomposition catalytic activity. *Dalton Transactions (Cambridge, England)*, (45): 6435–6440. doi:10.1039/b809923e PMID:19002331

Debecker, D. P., Stoyanova, M., Colbeau-Justin, F., Rodemerck, U., Boissière, C., Gaigneaux, E. M., & Sanchez, C. (2012). One-Pot Aerosol Route to MoO_3-SiO_2-Al_2O_3 Catalysts with Ordered Super Microporosity and High Olefin Metathesis Activity. *Angewandte Chemie International Edition, 51*(9), 2129–2131. doi:10.1002/anie.201106277 PMID:22262487

Deng, J., Ren, P., Deng, D., & Bao, X. (2015). Enhanced Electron Penetration through an Ultrathin Graphene Layer for Highly Efficient Catalysis of the Hydrogen Evolution Reaction. *Angewandte Chemie International Edition, 54*(7), 2100–2104. doi:10.1002/anie.201409524 PMID:25565666

Drochner, A., Kampe, P., Menning, N., Blickhan, N., Jekewitz, T., & Vogel, H. (2014). Acrolein Oxidation to Acrylic Acid on Mo/V/W-Mixed Oxide Catalysts. *Chemical Engineering & Technology, 37*(3), 398–408. doi:10.1002/ceat.201300797

Fan, X., Zhang, H., Li, J., Zhao, Z., Xu, C., Liu, J., & Wei, Y. et al. (2014). Ni-Mo nitride catalysts: Synthesis and application in the ammoxidation of propane. *Chinese Journal of Catalysis, 35*(3), 286–293. doi:10.1016/S1872-2067(14)60015-2

Fernandes, C. I., Capelli, S. C., Vaz, P. D., & Nunes, C. D. (2015). Highly selective and recyclable MoO_3 nanoparticles in epoxidation catalysis. *Applied Catalysis A, General, 504*, 344–350. doi:10.1016/j.apcata.2015.02.027

Frauwallner, M.-L., López-Linares, F., Lara-Romero, J., Scott, C. E., Ali, V., Hernández, E., & Pereira-Almao, P. (2011). Toluene hydrogenation at low temperature using a molybdenum carbide catalyst. *Applied Catalysis A, General, 394*(1-2), 62–70. doi:10.1016/j.apcata.2010.12.024

Fu, H., Qin, C., Lu, Y., Zhang, Z.-M., Li, Y.-G., Su, Z.-M., & Wang, E.-B. et al. (2012). An Ionothermal Synthetic Approach to Porous Polyoxometalate-Based Metal-Organic Frameworks. *Angewandte Chemie, 124*(32), 8109–8113. doi:10.1002/ange.201202994 PMID:22865560

Gaertner, C. A., van Veen, A. C., & Lercher, J. A. (2013). Oxidative Dehydrogenation of Ethane: Common Principles and Mechanistic Aspects. *ChemCatChem, 5*(11), 3196–3217. doi:10.1002/cctc.201200966

Gao, Q., Zhang, C., Xie, S., Hua, W., Zhang, Y., Ren, N., & Tang, Y. et al. (2009). Synthesis of Nanoporous Molybdenum Carbide Nanowires Based on Organic–Inorganic Hybrid Nanocomposites with Sub-Nanometer Periodic Structures. *Chemistry of Materials, 21*(23), 5560–5562. doi:10.1021/cm9014578

Giebeler, L., Wirth, A., Martens, J. A., Vogel, H., & Fuess, H. (2010). Phase transitions of V-Mo-W mixed oxides during reduction/re-oxidation cycles. *Applied Catalysis A, General, 379*(1–2), 155–165. doi:10.1016/j.apcata.2010.03.022

Giordano, C., Erpen, C., Yao, W., & Antonietti, M. (2008). Synthesis of Mo and W Carbide and Nitride Nanoparticles via a Simple Urea Glass Route. *Nano Letters, 8*(12), 4659–4663. doi:10.1021/nl8018593 PMID:19367981

Giordano, C., Erpen, C., Yao, W., Milke, B., & Antonietti, M. (2009). Metal Nitride and Metal Carbide Nanoparticles by a Soft Urea Pathway. *Chemistry of Materials*, *21*(21), 5136–5144. doi:10.1021/cm9018953

Griboval-Constant, A. (2004). Catalytic behaviour of cobalt or ruthenium supported molybdenum carbide catalysts for FT reaction. *Applied Catalysis A, General*, *260*(1), 35–45. doi:10.1016/j.apcata.2003.10.031

Guo, J., Li, Y., Zhu, S., Chen, Z., Liu, Q., Zhang, D., & Song, D.-M. et al. (2012). Synthesis of WO3@Graphene composite for enhanced photocatalytic oxygen evolution from water. *RSC Advances*, *2*(4), 1356–1363. doi:10.1039/C1RA00621E

Guzmán, H. J., Xu, W., Stacchiola, D., Vitale, G., Scott, C. E., Rodríguez, J. A., & Pereira-Almao, P. (2013). In situ time-resolved X-ray diffraction study of the synthesis of Mo2C with different carburization agents. *Canadian Journal of Chemistry*, *91*(7), 573–582. doi:10.1139/cjc-2012-0516

Guzmán, H. J., Xu, W., Stacchiola, D., Vitale, G., Scott, C. E., Rodríguez, J. A., & Pereira-Almao, P. (2015). Formation of β-Mo_2C below 600°C using MoO_2 nanoparticles as precursor. *Journal of Catalysis*, *332*, 83–94. doi:10.1016/j.jcat.2015.09.013

Han, J., Duan, J., Chen, P., Lou, H., Zheng, X., & Hong, H. (2012). Carbon-Supported Molybdenum Carbide Catalysts for the Conversion of Vegetable Oils. *ChemSusChem*, *5*(4), 727–733. doi:10.1002/cssc.201100476 PMID:22374620

Hao, X.-L., Ma, Y.-Y., Zang, H.-Y., Wang, Y.-H., Li, Y.-G., & Wang, E.-B. (2015). A Polyoxometalate-Encapsulating Cationic Metal-Organic Framework as a Heterogeneous Catalyst for Desulfurization. *Chemistry (Weinheim an der Bergstrasse, Germany)*, *21*(9), 3778–3784. doi:10.1002/chem.201405825 PMID:25612308

He, Q., Woo, J., Belianinov, A., Guliants, V. V., & Borisevich, A. Y. (2015). Better Catalysts through Microscopy: Mesoscale M1/M2 Intergrowth in Molybdenum-Vanadium Based Complex Oxide Catalysts for Propane Ammoxidation. *ACS Nano*, *9*(4), 3470–3478. doi:10.1021/acsnano.5b00271 PMID:25744246

Herrera, J. E., Kwak, J. H., Hu, J. Z., Wang, Y., Peden, C. H. F., Macht, J., & Iglesia, E. (2006). Synthesis, characterization, and catalytic function of novel highly dispersed tungsten oxide catalysts on mesoporous silica. *Journal of Catalysis*, *239*(1), 200–211. doi:10.1016/j.jcat.2006.01.034

Hollak, S. A. W., Gosselink, R. W., van Es, D. S., & Bitter, J. H. (2013). Comparison of Tungsten and Molybdenum Carbide Catalysts for the Hydrodeoxygenation of Oleic Acid. *ACS Catalysis*, *3*(12), 2837–2844. doi:10.1021/cs400744y

Holmberg, J., Hansen, S., Grasselli, R. K., & Andersson, A. (2006). Catalytic effects in propene ammoxidation achieved through substitutions in the M2 phase of the Mo-V-Nb-Te-oxide system. *Topics in Catalysis*, *38*(1–3), 17–29. doi:10.1007/s11244-006-0067-9

Hong, S. J., Lee, S., Jang, J. S., & Lee, J. S. (2011). Heterojunction BiVO4/WO3 electrodes for enhanced photoactivity of water oxidation. *Energy & Environmental Science*, *4*(5), 1781–1787. doi:10.1039/c0ee00743a

House, M. P., Carley, A. F., & Bowker, M. (2007). Selective oxidation of methanol on iron molybdate catalysts and the effects of surface reduction. *Journal of Catalysis*, *252*(1), 88–96. doi:10.1016/j.jcat.2007.09.005

Jaggers, C. H., Michaels, J. N., & Stacy, A. M. (1990). Preparation of high-surface-area transition-metal nitrides: Molybdenum nitrides, Mo2N and MoN. *Chemistry of Materials*, *2*(2), 150–157. doi:10.1021/cm00008a015

Jekewitz, T., Blickhan, N., Endres, S., Drochner, A., & Vogel, H. (2012). The influence of water on the selective oxidation of acrolein to acrylic acid on Mo/V/W-mixed oxides. *Catalysis Communications*, *20*, 25–28. doi:10.1016/j.catcom.2011.12.022

Jiao, Z., Wang, J., Ke, L., Sun, X. W., & Demir, H. V. (2011). Morphology-Tailored Synthesis of Tungsten Trioxide (Hydrate) Thin Films and Their Photocatalytic Properties. *ACS Applied Materials & Interfaces*, *3*(2), 229–236. doi:10.1021/am100875z PMID:21218846

Jones, S. D., Neal, L. M., & Hagelin-Weaver, H. E. (2008). Steam reforming of methanol using Cu-ZnO catalysts supported on nanoparticle alumina. *Applied Catalysis B: Environmental*, *84*(3–4), 631–642. doi:10.1016/j.apcatb.2008.05.023

Karim, K., & Sheikh, K. (2000). WO 200000284

Kim, T., Burrows, A., Kiely, C. J., & Wachs, I. E. (2007). Molecular/electronic structure-surface acidity relationships of model-supported tungsten oxide catalysts. *Journal of Catalysis*, *246*(2), 370–381. doi:10.1016/j.jcat.2006.12.018

Kitano, T., Okazaki, S., Shishido, T., Teramura, K., & Tanaka, T. (2013). Brønsted acid generation of alumina-supported molybdenum oxide calcined at high temperatures: Characterization by acid-catalyzed reactions and spectroscopic methods. *Journal of Molecular Catalysis A Chemical, 371*, 21–28. doi:10.1016/j.molcata.2013.01.019

Koós, Á., Barthos, R., & Solymosi, F. (2008). Reforming of Methanol on a K-Promoted Mo_2C/Norit Catalyst. *The Journal of Physical Chemistry C, 112*(7), 2607–2612. doi:10.1021/jp710015d

Kuang, X., Wu, X., Yu, R., Donahue, J. P., Huang, J., & Lu, C.-Z. (2010). Assembly of a metal–organic framework by sextuple intercatenation of discrete adamantane-like cages. *Nature Chemistry, 2*(6), 461–465. doi:10.1038/nchem.618 PMID:20489714

Kumar, A., Ali, A., Vinod, K. N., Mondal, A. K., Hegde, H., Menon, A., & Thimmappa, B. H. S. (2013). WO_x/ZrO_2: A highly efficient catalyst for alkylation of catechol with tert-butyl alcohol. *Journal of Molecular Catalysis A Chemical, 378*, 22–29. doi:10.1016/j.molcata.2013.05.010

Kunert, J., Drochner, A., Ott, J., Vogel, H., & Fueß, H. (2004). Synthesis of Mo/V mixed oxide catalysts via crystallisation and spray drying—a novel approach for controlled preparation of acrolein to acrylic acid catalysts. *Applied Catalysis A, General, 269*(1–2), 53–61. doi:10.1016/j.apcata.2004.03.050

Kurosaka, T., Maruyama, H., Naribayashi, I., & Sasaki, Y. (2008). Production of 1,3-propanediol by hydrogenolysis of glycerol catalyzed by Pt/WO_3/ZrO_2. *Catalysis Communications, 9*(6), 1360–1363. doi:10.1016/j.catcom.2007.11.034

Lausche, A. C., Schaidle, J. A., & Thompson, L. T. (2011). Understanding the effects of sulfur on Mo_2C and Pt/Mo_2C catalysts: Methanol steam reforming. *Applied Catalysis A, General, 401*(1–2), 29–36. doi:10.1016/j.apcata.2011.04.037

Li, X., & Iglesia, E. (2007). Synergistic effects of TiO2 and palladium-based cocatalysts on the selective oxidation of ethene to acetic acid on Mo-V-Nb oxide domains. *Angewandte Chemie International Edition, 46*(45), 8649–8652. doi:10.1002/anie.200700593 PMID:17926310

Li, X., & Iglesia, E. (2007). Synergistic effects of TiO2 and palladium-based cocatalysts on the selective oxidation of ethene to acetic acid on Mo-V-Nb oxide domains. *Angewandte Chemie International Edition, 46*(45), 8649–8652. doi:10.1002/anie.200700593 PMID:17926310

Li, X., & Zhang, Y. (2016). Highly Efficient Process for the Conversion of Glycerol to Acrylic Acid via Gas Phase Catalytic Oxidation of an Allyl Alcohol Intermediate. *ACS Catalysis*, *6*(1), 143–150. doi:10.1021/acscatal.5b01843

Liakakou, E. T., & Heracleous, E. (2016). Transition metal promoted K/Mo_2C as efficient catalysts for CO hydrogenation to higher alcohols. *Catal. Sci. Technol.*, *6*(4), 1106–1119. doi:10.1039/C5CY01173F

Liakakou, E. T., Heracleous, E., Triantafyllidis, K. S., & Lemonidou, A. A. (2015). K-promoted NiMo catalysts supported on activated carbon for the hydrogenation reaction of CO to higher alcohols: Effect of support and active metal. *Applied Catalysis B: Environmental*, *165*, 296–305. doi:10.1016/j.apcatb.2014.10.027

Liao, L., Wang, S., Xiao, J., Bian, X., Zhang, Y., Scanlon, M. D., & Girault, H. et al. (2014). A nanoporous molybdenum carbide nanowire as an electrocatalyst for hydrogen evolution reaction. *Energy Environmental Science*, *7*(1), 387–392. doi:10.1039/C3EE42441C

Lin, H., Liu, N., Shi, Z., Guo, Y., Tang, Y., & Gao, Q. (2016b). Cobalt-Doping in Molybdenum-Carbide Nanowires Toward Efficient Electrocatalytic Hydrogen Evolution. *Advanced Functional Materials*, *26*(31), 5590–5598. doi:10.1002/adfm.201600915

Lin, H., Shi, Z., He, S., Yu, X., Wang, S., Gao, Q., & Tang, Y. (2016a). Heteronanowires of MoC–Mo_2C as efficient electrocatalysts for hydrogen evolution reaction. *Chem. Sci.*, *7*(5), 3399–3405. doi:10.1039/C6SC00077K

Lin, M. M. (2001). Selective oxidation of propane to acrylic acid with molecular oxygen. *Applied Catalysis A, General*, *207*(1–2), 1–16. doi:10.1016/S0926-860X(00)00609-8

Liu, Q., Li, J., Zhao, Z., Gao, M., Kong, L., Liu, J., & Wei, Y. (2016). Synthesis, characterization, and catalytic performances of potassium-modified molybdenum-incorporated KIT-6 mesoporous silica catalysts for the selective oxidation of propane to acrolein. *Journal of Catalysis*, *344*, 38–52. doi:10.1016/j.jcat.2016.08.014

Liu, R., Lin, Y., Chou, L.-Y., Sheehan, S. W., He, W., Zhang, F., ... Wang, D. (2011). Water Splitting by Tungsten Oxide Prepared by Atomic Layer Deposition and Decorated with an Oxygen-Evolving Catalyst. Angewandte Chemie-International Edition, 50(2), 499–502.

Liu, Y., Ding, J., Sun, J., Zhang, J., Bi, J., Liu, K., & Chen, J. et al. (2016). Molybdenum carbide as an efficient catalyst for low-temperature hydrogenation of dimethyl oxalate. *Chemical Communications (Cambridge)*, 52(28), 5030–5032. doi:10.1039/C6CC01709F PMID:26983560

Liu, Y., Yu, G., Li, G.-D., Sun, Y., Asefa, T., Chen, W., & Zou, X. (2015). Coupling Mo_2C with Nitrogen-Rich Nanocarbon Leads to Efficient Hydrogen-Evolution Electrocatalytic Sites. *Angewandte Chemie International Edition*, 54(37), 10752–10757. doi:10.1002/anie.201504376 PMID:26212796

Lopez-Medina, R., Rojas, E., Banares, M. A., & Guerrero-Perez, M. O. (2012). Highly active and selective supported bulk nanostructured MoVNbTeO catalysts for the propane ammoxidation process. *Catalysis Today*, 192(1), 67–71. doi:10.1016/j.cattod.2011.11.018

Lunkenbein, T., Rosenthal, D., Otremba, T., Girgsdies, F., Li, Z., Sai, H., ... Breu, J. (2012). Access to Ordered Porous Molybdenum Oxycarbide/Carbon Nanocomposites. Angewandte Chemie International Edition, 51(51), 12892–12896.

Lwin, S., Li, Y., Frenkel, A. I., & Wachs, I. E. (2016). Nature of WO_x Sites on SiO_2 and Their Molecular Structure–Reactivity/Selectivity Relationships for Propylene Metathesis. *ACS Catalysis*, 6(5), 3061–3071. doi:10.1021/acscatal.6b00389

Ma, L., Ting, L. R. L., Molinari, V., Giordano, C., & Yei, B. S. (2015). Efficient hydrogen evolution reaction catalyzed by molybdenum carbide and molybdenum nitride nanocatalysts synthesized via the urea glass route. *Journal of Materials Chemistry A*, 3(16), 8361–8368. doi:10.1039/C5TA00139K

Ma, R., Hao, W., Ma, X., Tian, Y., & Li, Y. (2014). Catalytic Ethanolysis of Kraft Lignin into High-Value Small-Molecular Chemicals over a Nanostructured α-Molybdenum Carbide Catalyst. *Angewandte Chemie International Edition*, 126(28), 7438–7443. doi:10.1002/ange.201402752 PMID:24891069

Ma, Y., Guan, G., Phanthong, P., Hao, X., Huang, W., Tsutsumi, A., & Abudula, A. et al. (2014b). Catalytic Activity and Stability of Nickel-Modified Molybdenum Carbide Catalysts for Steam Reforming of Methanol. *The Journal of Physical Chemistry C*, 118(18), 9485–9496. doi:10.1021/jp501021t

Ma, Y., Guan, G., Shi, C., Zhu, A., Hao, X., Wang, Z., & Abudula, A. et al. (2014a). Low-temperature steam reforming of methanol to produce hydrogen over various metal-doped molybdenum carbide catalysts. *International Journal of Hydrogen Energy*, 39(1), 258–266. doi:10.1016/j.ijhydene.2013.09.150

McCrory, C. C. L., Jung, S., Ferrer, I. M., Chatman, S. M., Peters, J. C., & Jaramillo, T. F. (2015). Benchmarking Hydrogen Evolving Reaction and Oxygen Evolving Reaction Electrocatalysts for Solar Water Splitting Devices. *Journal of the American Chemical Society, 137*(13), 4347–4357. doi:10.1021/ja510442p PMID:25668483

McEnaney, J. M., Crompton, J. C., Callejas, J. F., Popczun, E. J., Biacchi, A. J., Lewis, N. S., & Schaak, R. E. (2014). Amorphous Molybdenum Phosphide Nanoparticles for Electrocatalytic Hydrogen Evolution. *Chemistry of Materials, 26*(16), 4826–4831. doi:10.1021/cm502035s

McKone, J. R., Sadtler, B. F., Werlang, C. A., Lewis, N. S., & Gray, H. B. (2013). Ni-Mo Nanopowders for Efficient Electrochemical Hydrogen Evolution. *ACS Catalysis, 3*(2), 166–169. doi:10.1021/cs300691m

Melzer, D., Xu, P., Hartmann, D., Zhu, Y., Browning, N. D., Sanchez-Sanchez, M., & Lercher, J. A. (2016). Atomic-Scale Determination of Active Facets on the MoVTeNb Oxide M1 Phase and Their Intrinsic Catalytic Activity for Ethane Oxidative Dehydrogenation. *Angewandte Chemie, 128*(31), 9019–9023. doi:10.1002/ange.201600463 PMID:26990594

Morales-Guio, C. G., & Hu, X. (2014). Amorphous Molybdenum Sulfides as Hydrogen Evolution Catalysts. *Accounts of Chemical Research, 47*(8), 2671–2681. doi:10.1021/ar5002022 PMID:25065612

Mouat, A. R., Lohr, T. L., Wegener, E. C., Miller, J. T., Delferro, M., Stair, P. C., & Marks, T. J. (2016). Reactivity of a Carbon-Supported Single-Site Molybdenum Dioxo Catalyst for Biodiesel Synthesis. *ACS Catalysis, 6*(10), 6762–6769. doi:10.1021/acscatal.6b01717

Nishiguichi, T., Oka, K., Matsumoto, T., Kanai, H., Utani, K., & Imamura, S. (2006). Durability of WO_3/ZrO_2-CuO/CeO_2 catalysts for steam reforming of dimethyl ether. *Applied Catalysis A, General, 301*(1), 66–74. doi:10.1016/j.apcata.2005.11.011

Noh, H., Cui, Y., Peters, A. W., Pahls, D. R., Ortuño, M. A., Vermeulen, N. A., & Farha, O. K. et al. (2016). An Exceptionally Stable Metal–Organic Framework Supported Molybdenum(VI) Oxide Catalyst for Cyclohexene Epoxidation. *Journal of the American Chemical Society, 138*(44), 14720–14726. doi:10.1021/jacs.6b08898 PMID:27779867

Pan, L. F., Li, Y. H., Yang, S., Liu, P. F., Yu, M. Q., & Yang, H. G. (2014). Molybdenum carbide stabilized on graphene with high electrocatalytic activity for hydrogen evolution reaction. *Chemical Communications*, *50*(86), 13135–13137. doi:10.1039/C4CC05698A PMID:25229076

Pang, Y., Chen, X., Xu, C., Lei, Y., & Wei, K. (2014). High Catalytic Performance of MoO_3-Bi_2SiO_5/SiO_2 for the Gas-Phase Epoxidation of Propylene by Molecular Oxygen. *ChemCatChem*, *6*(3), 876–884. doi:10.1002/cctc.201300811

Park, J. H., Kim, Y. T., Park, E. D., Lee, H. C., Kim, J., & Lee, D. (2013). Markedly High Catalytic Activity of Supported $PtMoO_x$ Nanoclusters for Methanol Reforming to Hydrogen at Low Temperatures. *ChemCatChem*, *5*(3), 806–814. doi:10.1002/cctc.201200458

Phuruangrat, A., Ham, D. J., Hong, S. J., Thongtem, S., & Lee, J. S. (2010). Synthesis of hexagonal WO_3 nanowires by microwave-assisted hydrothermal method and their electrocatalytic activities for hydrogen evolution reaction. *Journal of Materials Chemistry*, *20*(9), 1683–1690. doi:10.1039/B918783A

Posada-Pérez, S., Ramírez, P. J., Evans, J., Viñes, F., Liu, P., Illas, F., & Rodriguez, J. A. (2016). Highly Active Au/δ-MoC and Cu/δ-MoC Catalysts for the Conversion of CO_2 : The Metal/C Ratio as a Key Factor Defining Activity, Selectivity, and Stability. *Journal of the American Chemical Society*, *138*(26), 8269–8278. doi:10.1021/jacs.6b04529 PMID:27308923

Prins, R., & Bussell, M. E. (2012). Metal Phosphides: Preparation, Characterization and Catalytic Reactivity. *Catalysis Letters*, *142*(12), 1413–1436. doi:10.1007/s10562-012-0929-7

Ross-Medgaarden, E. I., Knowles, W. V., Kim, T., Wong, M. S., Zhou, W., Kiely, C. J., & Wachs, I. E. (2008). New insights into the nature of the acidic catalytic active sites present in ZrO_2-supported tungsten oxide catalysts. *Journal of Catalysis*, *256*(1), 108–125. doi:10.1016/j.jcat.2008.03.003

Sabnis, K. D., Akatay, M. C., Cui, Y., Sollberger, F. G., Stach, E. A., Miller, J. T., & Ribeiro, F. H. et al. (2015a). Probing the active sites for water–gas shift over Pt/molybdenum carbide using multi-walled carbon nanotubes. *Journal of Catalysis*, *330*, 442–451. doi:10.1016/j.jcat.2015.07.032

Sabnis, K. D., Cui, Y., Akatay, M. C., Shekhar, M., Lee, W.-S., Miller, J. T., & Ribeiro, F. H. et al. (2015b). Water–gas shift catalysis over transition metals supported on molybdenum carbide. *Journal of Catalysis*, *331*, 162–171. doi:10.1016/j.jcat.2015.08.017

Sadakane, M., Watanabe, N., Katou, T., Nodasaka, Y., & Ueda, W. (2007). Crystalline Mo_3VO_x mixed-metal-oxide catalyst with trigonal symmetry. *Angewandte Chemie International Edition, 46*(9), 1493–1496. doi:10.1002/anie.200603923 PMID:17221899

Schaidle, J. A., Lausche, A. C., & Thompson, L. T. (2010). Effects of sulfur on Mo_2C and Pt/Mo_2C catalysts: Water gas shift reaction. *Journal of Catalysis, 272*(2), 235–245. doi:10.1016/j.jcat.2010.04.004

Schaidle, J. A., Schweitzer, N. M., Ajenifujah, O. T., & Thompson, L. T. (2012). On the preparation of molybdenum carbide-supported metal catalysts. *Journal of Catalysis, 289*, 210–217. doi:10.1016/j.jcat.2012.02.012

Schweitzer, N. M., Schaidle, J. A., Ezekoye, O. K., Pan, X., Linic, S., & Thompson, L. T. (2011). High Activity Carbide Supported Catalysts for Water Gas Shift. *Journal of the American Chemical Society, 133*(8), 2378–2381. doi:10.1021/ja110705a PMID:21291250

Setthapun, W., Bej, S. K., & Thompson, L. T. (2008). Carbide and Nitride Supported Methanol Steam Reforming Catalysts: Parallel Synthesis and High Throughput Screening. *Topics in Catalysis, 49*(1–2), 73–80. doi:10.1007/s11244-008-9070-7

Shi, X., Choi, Y., Zhang, K., Kwon, J., Kim, D. Y., Lee, J. K., & Park, J. H. et al. (2014). Efficient photoelectrochemical hydrogen production from bismuth vanadate-decorated tungsten trioxide helix nanostructures. *Nature Communications, 5*, 4775. doi:10.1038/ncomms5775 PMID:25179126

Shiju, N. R., & Guliants, V. V. (2007). Microwave-assisted hydrothermal synthesis of monophasic Mo-V-Te-Nb-O mixed oxide catalyst for the selective ammoxidation of propane. *ChemPhysChem, 8*(11), 1615–1617. doi:10.1002/cphc.200700257 PMID:17614349

Shou, H., Ferrari, D., Barton, D. G., Jones, C. W., & Davis, R. J. (2012). Influence of Passivation on the Reactivity of Unpromoted and Rb-Promoted Mo2C Nanoparticles for CO Hydrogenation. *ACS Catalysis, 2*(7), 1408–1416. doi:10.1021/cs300083b

Sobolev, V. I., & Koltunov, K. Y. (2011). MoVNbTe Mixed Oxides as Efficient Catalyst for Selective Oxidation of Ethanol to Acetic Acid. *ChemCatChem, 3*(7), 1143–1145. doi:10.1002/cctc.201000450

Solsona, B., López Nieto, J., Oliver, J., & Gumbau, J. (2004). Selective oxidation of propane and propene on MoVNbTeO catalysts. *Catalysis Today, 91–92*, 247–250. doi:10.1016/j.cattod.2004.03.058

Song, K., Zhang, H., Zhang, Y., Tang, Y., & Tang, K. (2013). Preparation and characterization of WO_x/ZrO_2 nanosized catalysts with high WO_x dispersion threshold and acidity. *Journal of Catalysis, 299*, 119–128. doi:10.1016/j.jcat.2012.11.011

Spamer, A., Dube, T., Moodley, D., van Schalkwyk, C., & Botha, J. (2003). The reduction of isomerisation activity on a WO_3/SiO_2 metathesis catalyst. *Applied Catalysis A, General, 255*(2), 153–167. doi:10.1016/S0926-860X(03)00537-4

Széchenyi, A., & Solymosi, F. (2007). Production of Hydrogen in the Decomposition of Ethanol and Methanol over Unsupported Mo_2C Catalysts. *The Journal of Physical Chemistry C, 111*(26), 9509–9515. doi:10.1021/jp072439k

Tagusagawa, C., Takagaki, A., Takanabe, K., Ebitani, K., Hayashi, S., & Domen, K. (2009). Effects of Transition-Metal Composition of Protonated, Layered Nonstoichiometric Oxides $H_{1-x}Nb_{1-x}Mo_{1+x}O_6$ on Heterogeneous Acid Catalysis. *The Journal of Physical Chemistry C, 113*(40), 17421–17427. doi:10.1021/jp906628k

Triwahyono, S., Jalil, A. A., Ruslan, N. N., Setiabudi, H. D., & Kamarudin, N. H. N. (2013). C5–C7 linear alkane hydroisomerization over MoO_3-ZrO2 and Pt/MoO_3-ZrO_2 catalysts. *Journal of Catalysis, 303*, 50–59. doi:10.1016/j.jcat.2013.03.016

Tsukuda, E., Sato, S., Takahashi, R., & Sodesawa, T. (2007). Production of acrolein from glycerol over silica-supported heteropoly acids. *Catalysis Communications, 8*(9), 1349–1353. doi:10.1016/j.catcom.2006.12.006

Tuomi, S., Guil-Lopez, R., & Kallio, T. (2016). Molybdenum carbide nanoparticles as a catalyst for the hydrogen evolution reaction and the effect of pH. *Journal of Catalysis, 334*, 102–109. doi:10.1016/j.jcat.2015.11.018

van der Klis, F., Le Nôtre, J., Blaauw, R., van Haveren, J., & van Es, D. S. (2012). Renewable linear alpha olefins by selective ethenolysis of decarboxylated unsaturated fatty acids. *European Journal of Lipid Science and Technology, 114*(8), 911–918. doi:10.1002/ejlt.201200024

Vrubel, H., & Hu, X. (2012). Molybdenum Boride and Carbide Catalyze Hydrogen Evolution in both Acidic and Basic Solutions. *Angewandte Chemie*, *124*(51), 12875–12878. doi:10.1002/ange.201207111 PMID:23143996

Wan, C., Regmi, Y. N., & Leonard, B. M. (2014). Multiple Phases of Molybdenum Carbide as Electrocatalysts for the Hydrogen Evolution Reaction. *Angewandte Chemie*, *126*(25), 6525–6528. doi:10.1002/ange.201402998 PMID:24827779

Wang, F., & Ueda, W. (2009). Selective oxidation of alcohols using novel crystalline Mo-V-O oxide as heterogeneous catalyst in liquid phase with molecular oxygen. *Catalysis Today*, *144*(3–4), 358–361. doi:10.1016/j.cattod.2008.12.034

Wang, G., Ling, Y., Wang, H., Yang, X., Wang, C., Zhang, J. Z., & Li, Y. (2012). Hydrogen-treated WO_3 nanoflakes show enhanced photostability. *Energy & Environmental Science*, *5*(3), 6180–6187. doi:10.1039/c2ee03158b

Wang, S., Antonio, D., Yu, X., Zhang, J., Cornelius, A. L., He, D., & Zhao, Y. (2015). The hardest superconducting metal nitride. *Scientific Reports*, *5*, 13733–13741. doi:10.1038/srep13733 PMID:26333418

Wang, S., Ge, H., Sun, S., Zhang, J., Liu, F., Wen, D., & Zhao, Y. et al. (2015). New molybdenum nitride catalyst with rhombohedral MoS_2 structure for hydrogenation applications. *Journal of the American Chemical Society*, *137*(14), 4815–4822. doi:10.1021/jacs.5b01446 PMID:25799018

Wang, S., Yu, X., Lin, Z., Zhang, R., He, D., Qin, J., & Zhao, Y. et al. (2012). Synthesis, crystal structure, and elastic properties of novel tungsten nitrides. *Chemistry of Materials*, *24*(15), 3023–3028. doi:10.1021/cm301516w

Wang, T., Wang, X., Liu, Y., Zheng, J., & Li, X. (2016). A highly efficient and stable biphasic nanocrystalline Ni-Mo-N catalyst for hydrogen evolution in both acidic and alkaline electrolytes. *Nano Energy*, *22*, 111–119. doi:10.1016/j.nanoen.2016.02.023

Wang, Z., Ma, Y., Zhang, M., Li, W., & Tao, K. (2008). A novel route to the synthesis of bulk and well dispersed alumina-supported Ni_2Mo_3N catalysts via single-step hydrogen thermal treatment. *Journal of Materials Chemistry*, *18*(37), 4421–4425. doi:10.1039/b807748g

Wei, M., He, C., Hua, W., Duan, C., Li, S., & Meng, Q. (2006). A Large Protonated Water Cluster H + $(H_2O)_{27}$ in a 3D Metal−Organic Framework. *Journal of the American Chemical Society, 128*(41), 13318–13319. doi:10.1021/ja0611184 PMID:17031919

Wu, H. B., Xia, B. Y., Yu, L., Yu, X.-Y., & Lou, X. W. (2015). Porous molybdenum carbide nano-octahedrons synthesized via confined carburization in metal-organic frameworks for efficient hydrogen production. *Nature Communications, 6*, 6512. doi:10.1038/ncomms7512 PMID:25758159

Wu, Z., Yang, Y., Gu, D., Li, Q., Feng, D., Chen, Z., & Zhao, D. et al. (2009). Silica-Templated Synthesis of Ordered Mesoporous Tungsten Carbide/Graphitic Carbon Composites with Nanocrystalline Walls and High Surface Areas via a Temperature-Programmed Carburization Route. *Small, 5*(23), 2738–2749. doi:10.1002/smll.200900523 PMID:19743431

Xi, G., Ouyang, S., Li, P., Ye, J., Ma, Q., Su, N., ... Wang, C. (2012). Ultrathin W18O49 Nanowires with Diameters below 1 nm: Synthesis, Near-Infrared Absorption, Photoluminescence, and Photochemical Reduction of Carbon Dioxide. Angewandte Chemie-International Edition, 51(10), 2395–2399.

Xiao, P., Ge, X., Wang, H., Liu, Z., Fisher, A., & Wang, X. (2015). Novel Molybdenum Carbide-Tungsten Carbide Composite Nanowires and Their Electrochemical Activation for Efficient and Stable Hydrogen Evolution. *Advanced Functional Materials, 25*(10), 1520–1526. doi:10.1002/adfm.201403633

Xiao, P., Sk, M. A., Thia, L., Ge, X., Lim, R. J., Wang, J.-Y., & Wang, X. et al. (2014). Molybdenum phosphide as an efficient electrocatalyst for the hydrogen evolution reaction. *Energy Environ. Sci., 7*(8), 2624–2629. doi:10.1039/C4EE00957F

Xiao, T., York, A. P. E., Coleman, K. S., Claridge, J. B., Sloan, J., Charnock, J., & Green, M. L. H. (2001). Effect of carburising agent on the structure of molybdenum carbides. *Journal of Materials Chemistry A, 11*(12), 3094–3098. doi:10.1039/b104011c

Xie, Y. P., Liu, G., Yin, L., & Cheng, H.-M. (2012). Crystal facet-dependent photocatalytic oxidation and reduction reactivity of monoclinic WO_3 for solar energy conversion. *Journal of Materials Chemistry, 22*(14), 6746–6751. doi:10.1039/c2jm16178h

Xing, Z., Liu, Q., Asiri, A. M., & Sun, X. (2014). Closely Interconnected Network of Molybdenum Phosphide Nanoparticles: A Highly Efficient Electrocatalyst for Generating Hydrogen from Water. *Advanced Materials*, *26*(32), 5702–5707. doi:10.1002/adma.201401692 PMID:24956199

Yan, H., Jiao, Y., Wu, A., Tian, C., Zhang, X., Wang, L., & Fu, H. et al. (2016). Cluster-like molybdenum phosphide anchored on reduced graphene oxide for efficient hydrogen evolution over a broad pH range. *Chemical Communications*, *52*(61), 9530–9533. doi:10.1039/C6CC04220A PMID:27380736

Yang, J., Zhang, F., Wang, X., He, D., Wu, G., Yang, Q., & Li, Y. et al. (2016b). Porous Molybdenum Phosphide Nano-Octahedrons Derived from Confined Phosphorization in UIO-66 for Efficient Hydrogen Evolution. *Angewandte Chemie International Edition*, *55*(41), 12854–12858. doi:10.1002/anie.201604315 PMID:27384250

Yang, X., Feng, X., Tan, H., Zang, H., Wang, X., Wang, Y., & Li, Y. et al. (2016a). N-Doped graphene-coated molybdenum carbide nanoparticles as highly efficient electrocatalysts for the hydrogen evolution reaction. *Journal of Materials Chemistry A*, *4*(10), 3947–3954. doi:10.1039/C5TA09507G

Youn, D. H., Han, S., Kim, J. Y., Kim, J. Y., Park, H., Choi, S. H., & Lee, J. S. (2014). Highly Active and Stable Hydrogen Evolution Electrocatalysts Based on Molybdenum Compounds on Carbon Nanotube–Graphene Hybrid Support. *ACS Nano*, *8*(5), 5164–5173. doi:10.1021/nn5012144 PMID:24787540

Yu, R., Kuang, X.-F., Wu, X.-Y., Lu, C.-Z., & Donahue, J. P. (2009). Stabilization and immobilization of polyoxometalates in porous coordination polymers through host–guest interactions. *Coordination Chemistry Reviews*, *253*(23-24), 2872–2890. doi:10.1016/j.ccr.2009.07.003

Zhang, H., Han, J., Niu, X., Han, X., Wei, G., & Han, W. (2011). Study of synthesis and catalytic property of WO_3/TiO_2 catalysts for NO reduction at high temperatures. *Journal of Molecular Catalysis A Chemical*, *350*(1–2), 35–39. doi:10.1016/j.molcata.2011.09.001

Zhang, Q., Li, H., Liu, X., Qin, W., Zhang, Y., Xue, W., & Yang, S. (2013). Modified Porous Zr-Mo Mixed Oxides as Strong Acid Catalysts for Biodiesel Production. *Energy Technology*, *1*(12), 735–742. doi:10.1002/ente.201300109

Zhang, Z., Zhang, Q., Jia, L., Wang, W., Tian, S. P., Wang, P., & Tan, Y. et al. (2016b). The effects of the Mo–Sn contact interface on the oxidation reaction of dimethyl ether to methyl formate at a low reaction temperature. *Catalysis Science & Technology*, 6(15), 6109–6117. doi:10.1039/C6CY00460A

Zhang, Z., Zhang, Q., Jia, L., Wang, W., Xiao, H., Han, Y., & Tan, Y. et al. (2016a). Effects of MoO_3 crystalline structure of MoO_3-SnO_2 catalysts on selective oxidation of glycol dimethyl ether to 1,2-propandiol. *Catalysis Science & Technology*, 6(6), 1842–1849. doi:10.1039/C5CY00894H

Zhao, Z., Yang, H., Li, Y., & Guo, X. (2014). Cobalt-modified molybdenum carbide as an efficient catalyst for chemoselective reduction of aromatic nitro compounds. *Green Chemistry*, 16(3), 1274–1281. doi:10.1039/C3GC42049C

Zhou, W., Ross-Medgaarden, E. I., Knowles, W. V., Wong, M. S., Wachs, I. E., & Kiely, C. J. (2009). Identification of active Zr-WO_x clusters on a ZrO_2 support for solid acid catalysts. *Nature Chemistry*, 1(9), 722–728. doi:10.1038/nchem.433 PMID:21124359

Chapter 4
Low-Dimensional Molybdenum-Based Catalytic Materials from Theoretical Perspectives

The most intriguing development in chemistry and material science in the past few decades is perhaps the discovery of low-dimensional materials. The intensive studies of nanoparticles, or zero-dimensional (0D) during the 1970s and 80s give rise to a new research field, called "nanotechnology", which has been a hot subject since then, and have greatly affected the chemical industry. The later discovery of graphene and its amazing physical and chemical properties opened the gate of a new class of materials, namely, two-dimensional (2D) materials. Afterwards, enormous scientific and technological research efforts were invested in the development of new 2D materials, as well as the application of the materials in various fields, including catalysis, pharmaceutical industry, optics, electronics, mechanics, etc.

Due to the microscopic nature of the low-dimensional materials, the characterization and evaluation of the materials may miss some important information because of the complexity of the system and the inhomogeneity of the materials that are prepared. Over the past years, quantum chemical calculations, especially density functional theory (DFT) have emerged as a new tool to understand the structure of the catalysts and the mechanism of the catalytic reactions. From theoretical point of view, the unique properties of 0D nanoparticles 2D materials come from the 0D and 2D confinement of the materials, which strongly modifies their electronic structures. For example, the 0D or 2D form of a semiconductor or insulator can be metallic. The 0D or 3D form of a material can have different catalytic activity and selectivity in catalysis. The subject to be discussed in this section is the 0D molybdenum carbide (Mo_2C) nanoparticle and the 2D molybdenum sulfide (MoS_2) and molybdenum nitride (MoN_2), focusing on their structures, electronic structures and the mechanism of catalytic reactions that are obtained from theoretical perspectives.

1. MoS_2 2D AND 0D CATALYTIC MATERIALS

1.1 Structures and Electronic Structures

Bulk is MoS_2 is a lubricant and a catalyst material widely used in industry. It is known as a diamagnetic, indirect bandgap semiconductor(Kobayashi & Yamauchi, 1995). MoS_2 is different from other catalytic materials in that it has a graphite-like structure, where the interactions between the layers are weak van der Waals interactions in nature. After graphene was isolated from graphite, few-layer MoS_2 materials are also synthesized in large scale(Coleman et al., 2011). These materials are chemically active in many applications because of their special electronic structures compared to the bulk (Chhowalla et al. 2013).

The key to the understanding of low-dimensional material, compared to bulk, is the effect of quantum confinement on the electronic structure. With the decreasing of the number of layers, the indirect band gap in the near-infrared frequency range ($\Delta = 1.2$ eV) shifted up by about 0.7 eV to direct band gap in the range of the visible light ($\Delta = 1.2$ eV), and the indirect-gap bulk MoS_2 becomes a direct-gap monolayer MoS_2, which emits visible light strongly (Figure 1). This theoretical prediction is confirmed by optical spectroscopy studies (Mak et al. 2010).

Moreover, the electronic structure of MoS_2 changes significantly when it is exposed to external electric or magnetic field (Kormányos et al. 2014), or strain. External electric field induces intrinsic and the Bychkov-Rashba spin-orbit coupling in MoS_2, especially for bilayers. Under isotropic or uniaxial tensile strain (Lu et al. 2012), the direct band gap of MoS_2 monolayer changes back to an indirect band gap, as in bulk (Figure 2).

Figure 1. Band structures of MoS_2 with different number of layers.
Source: Reproduced with permission from Zahid et al. (2013); used in accordance with the Creative Commons Attribution (CC BY) license (https://creativecommons.org/licenses/by/4.0/)."

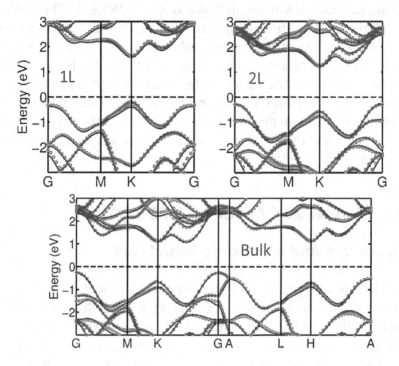

It is known that defects sites are in many cases the active sites for catalytic reactions. Like graphene, Monolayer MoS_2 have intrinsic point defects, grain boundaries, and edges (Zhou et al., 2013). Indeed, grain boundaries and edges have unique characteristics compared to the intact surface, which could potentially change their local catalytic activity. Band structure and local density of states (LDOS) shows that the grain boundary can serve as 1D conducting metallic quantum wires embedded in the semi-conducting MoS_2 plane, but the conductivity can be greatly undermined if multiple kinks exists on the grain boundary (Figure 3). Similar to the grain boundary, the edges maintains clear metallic nature, but with distinct magnetic properties. Every Mo atoms on unconstructed edges have a local magnetic moment of 0.4 μ_B, while the Mo magnetic moments are quenched in the reconstructed Mo edges. (Figure 4)

Recently, it has been reported that monolayer MoS_2 can also form heterostructures with other 2D transition metal dichalcogenides such as WS_2, leading to materials with new electronic structures (Gong et al. 2014). Interlayer Electrostatic Coulomb interactions between the electrons and electron

Figure 2. (a) Strain dependence of band gap energies of 1L-MoS2 (a = 3.160 A°). The representative band structures for the (b) compressive and (c), (d) tensile stresses are displayed, respectively. Inset indicates the hexagonal structure consisting of Mo (red/gray balls) and S (yellow/light gray balls) from the top views.
Source: Reproduced with permission from Yun et al. (2012). Copyright of American Physical Society.

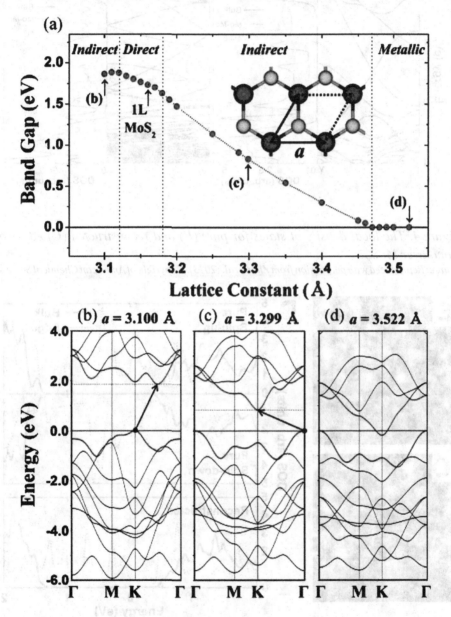

Figure 3. The band structure and local density of states (LDOS) of the clean grain boundarie (left) and grain boundary with kinks (right) of monolayer MoS$_2$.
Source: Reproduced with permission from Zhou et al. (2013). Copyright of American Chemical Society.

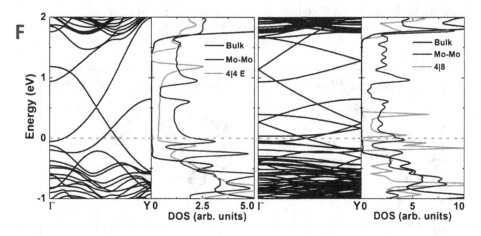

Figure 4. The local density of states for pure (B) and reconstructed (C) edges of monolayer MoS$_2$.
Source: Reproduced with permission from Zhou et al. (2013). Copyright of American Chemical Society.

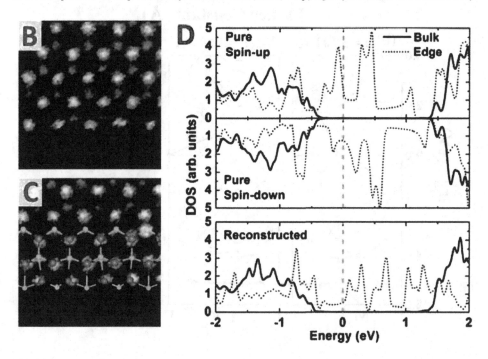

holes (excitonic interactions) exist between the MoS_2 and WS_2 layers, as indicated by photoluminescence spectroscopy. The electronic structure of the interlayers depends on the interface structure of WS_2/MoS_2. Zigzag WS_2/MoS_2 monolayer interface shows direct transitions at 1.825 eV, 1.875 eV, 1.889 eV, 1.908 eV, 1.968 eV, and 1.987 eV. The armchair WS_2/MoS_2 monolayer interface shows direct transitions at 1.835 eV, 1.925 eV, 1.939 eV, 1.978 eV, and 2.06 eV.

It is known that MoS_2 catalyst exists as dispersed nanosized monolayer particles in reactions such as HDS. However, the particle can have multiple shapes, such as triangular and hexagonal, etc. Also, the termination of the edge can have Mo edge and S edge with different S coverages at catalytic working conditions. Using DFT and Gibbs-Curie-Wulff theorem, the structure and stability of the MoS_2 single-layer clusters (S-Mo-S) were systematically studied. It was predicted that hexagonal shape is thermodynamically stable under HDS working conditions. However, at high H_2S partial pressure, triangular-shaped clusters, which were commonly observed in STM experiments (Helveg et al. 2000; Lauritsen et al. 2004), are the most stable morphology (Wen et al. 2005). Investigations on the electronic properties of a hexagonal $Mo_{27}S_{54}$ cluster (Orita et al. 2003a) show that the corner Mo atoms ($Mo_{C, IV}$) are more active than others, possessing the most positive charge. This is further confirm by the HOMO orbital diagram, where the HOMO and HOMO-1 orbitals are localized more on the corner Mo atoms. All the molecular orbitals near Fermi level are contributed dominantly by Mo 4d electrons, while the contribution from S atoms is very small.

Schweiger et al (Schweiger, Raybaud, Kresse, & Toulhoat, 2002) also predicted that the Mo edge is energetically the most stable surface under realistic HDS conditions. Wen et al. (2005) used cluster models with sizes close to real MoS_2 particles and studied the interaction between Mo and S edges (Figure 5) in MoS_2 nanoparticles. Calculations on the energetics of the formation of sulfur coverages show that the formation energy of sulfur coverages depends on the sulfur coverage of the precursor state. It is exothermic for 67% and 50% sulfur coverage, while endothermic for 33% and 0% sulfur coverage. At 675K and medium H_2S/H_2 ratio, two stable structures with 33% and 50% can coexist, while at very high H_2S/H_2 ratio, 100% sulfur coverage on both the Mo edge and S edge is possible, but the Mo edge with 100% sulfur coverage is more favorable than that the S edge with the same sulfur coverage.

Figure 5. The Mo and S edges with different S coverages. 100% sulfur coverage corresponding to two S atoms per Mo atom on the Mo edge, then adsorption of two, three, four, and six S atoms on the Mo edge of clusters, which gives 33, 50, 67, and 100% sulfur coverage, respectively. For the S edge, 100% sulfur coverage corresponds to six bridge S atoms. Consequently, two, three, and four S atoms on the S edge show 33, 50, and 67% sulfur coverage, respectively.
Source: Reproduced with permission from Wen et al. (2005). Copyright of American Chemical Society.

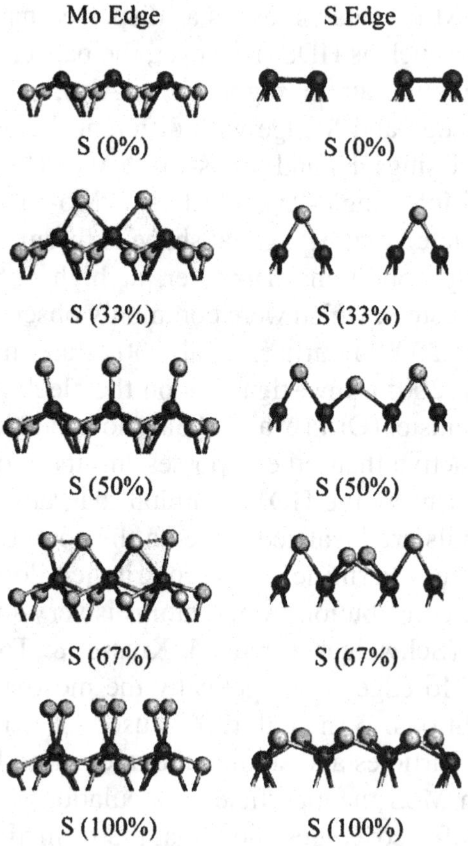

Mo Edge	S Edge
S (0%)	S (0%)
S (33%)	S (33%)
S (50%)	S (50%)
S (67%)	S (67%)
S (100%)	S (100%)

Besides the layered nanoclusters, MoS_2 can also form octahedron type fullerene structures (Figure 6), where nanocages of about 30 nm are formed by a single-wall of MoS_2 layer (Bar-Sadan et al. 2006). The experimentally obtained structure perfectly matches the structure obtained by quantum mechanical molecular dynamic simulations. Theoretical study shows that MoS_2 can form other structures, including nanotubes(Stefanov et al. 2008), nanoribbons(Xiao et al. 2016), etc., but these are outside the scope of this chapter.

Figure 6. TEM images of a three-layer nano-octahedron at various tilt angles (a) and corresponding views of the model structure of an $(MoS_2)784@(MoS_2)1296@(MoS_2)1936$ fullerene (b). Note the similarity between the experimentally observed and model structures.
Source: Reproduced with permission from Bar-Sadan et al. (2006). Copyright of American Chemical Society.

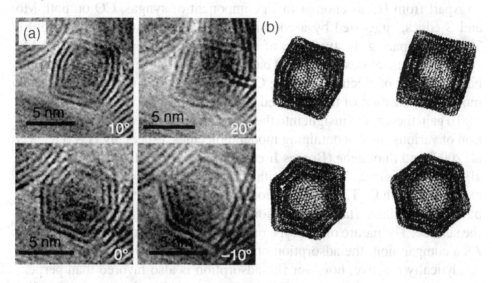

1.2 The Adsorption of Small Molecules on MoS_2 Nanoparticles

Since MoS_2 is mostly used as hydro-treating catalyst, the adsorption of hydrogen species on the surface of MoS_2 is crucial to the catalytic activity. Using cluster model DFT, Wen et al (Wen et al. 2006) have shown that molecular H_2 adsorption sites depends on the surface coverage. At low coverage, molecular H_2 tend to dissociate homolytically on the S edge. However at high coverage, the Mo sites are more favored thermodynamically than the S sites.

It has been proposed since 1990s that oxygen plays important role in the MoS_2 catalyzed HDS reactions. Chary et al. (1991) found a linear correlation between oxygen chemisorption on MoS_2 and the HDS activity. They thus propose that oxygen chemisorption on MoS_2 and HDS behaves in a similar manner, and that both happen on the coordinatively unsaturated sites (CUS). Indeed, DFT studies (KC et al. 2015 and Sen et al. 2014) on the adsorption of O_2 on MoS_2 planear surface shows that the molecular adsorption is very weak, and the dissociative adsorption is kinetically limited on the surface at low temperature due to large energy barriers. Despite of the weak adsorption on the surface, our DFT studies (Liu et al. 2016) show that oxygen can easily bind to the CUS sites, substitute the edge S atoms, leading to a oxidation state. By systematically study the successive oxidation of the $Mo_{27}S_xO_y$ nanoparticles, it can be concluded that the oxidation of MoS_2 is thermodynamically favorable, which agrees well with other studies (Fleischauer & Lince, 1999, KC et al. 2015 and Liang et al. 2011).

Apart from H_2, as another main component of syngas, CO on both Mo and S edges, suggested by a combined DFT and IR study (Travert et al. 2006). Comparing the two type of sites, the Mo edge with decreased Mo coordination increases the filled d band at the Fermi level, thus increases the back-donation of electrons into the CO $2\pi^*$ antibonding orbital, and leads to more destablization of the adsorbed CO molecule.

To gain theoretical insight into the first step of HDS reactions, the adsorption of various sulfur-containing model molecules, especially H_2S (Galea et al. 2009) and thiophene (Borges Jr et al. 2007, Cristol et al. 2004, Cristol et al. 2006, Moses et al. 2009 and Orita et al. 2003b) on MoS_2 has been studied extensively with DFT. The adsorption of thiophene and its derivatives on the basal planes takes a flat configuration, with a seperation of about 3.5 Å from the surface. The nature of this type of adsorption is wan der Waals in nature. As a comparision, the adsorption on the CUS of the edges are found to be catalytically reactive, however flat adsorption is also favored than perpendicular adsorption on the edges (Borges Jr et al. 2007 and Orita et al. 2003b). Hydrogenation increases the adsorption energy of these molecules, and makes them more likely to be adsorbed perpendicularly (Liu et al. 2016). After adsorption, the C-S bonds are activated, as indicated by the C-S distances. For larger sulfur-containning aromatic molecules such as dibenzothiophene (DBT) and its derivatives, steric effects plays important role. Our studies show that steric effect by flat adsorption is smaller than that of by perpendicular adsorption, the DBT derivatives such as 4,6-dimethyl-substituted dibenzothiophene (4,6-DMDBT) also favors flat adsorption. Competition effects exist among the sulfur-containnig molecules, which have recently been found by our studies (Liu et al. 2016) and Rangarajan & Mavrikakis (2016). Besides, nitrogen-containning molecule can also adsorb on the brim sites of the MoS_2 catalyst, and inhibit the HDS reactions on these sites, leaving edge vacancy sites as the most important sites.

1.3 The Mechanism of MoS_2 Catalyzed HDS Reactions

Thiophene is typically used as a model reactant for understand the mechanism of HDS. Thiophene HDS over MoS_2 catalysts has been studied experimentally by some groups. Its reaction mechanism has been proposed by Sullivan

Figure 7. The sum of mechanisms for thiophene HDS

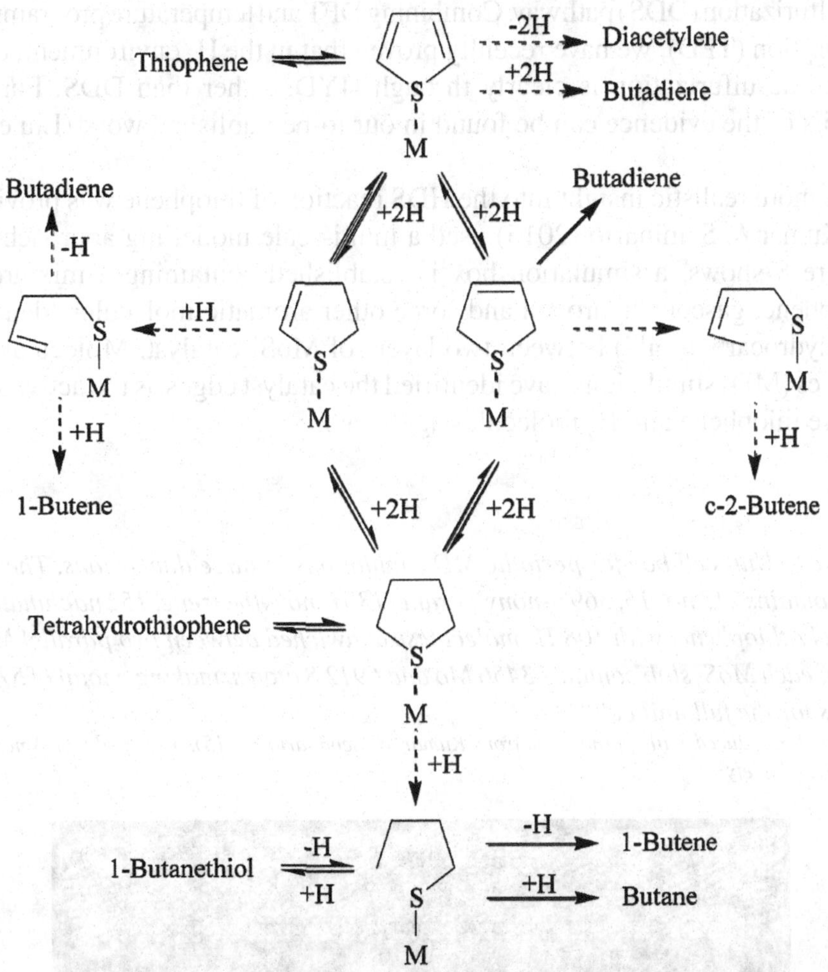

et al.(Sullivan & Ekerdt, 1998), which are summarized in Figure 7. Dihydrothiphene (DHT), tetrahydrothiophene (THT), and 1-butanethiolate are proposed intermediates in thiophene HDS.(Devanneaux & Maurin 1981, Hargreaves & Ross 1979, Hensen et al. 1996, Liu & Friend 1991, Markel et al. 1989, Roberts & Friend 1986 and Sullivan & Ekerdt 1998) Furthermore, the mechanisms of HDS for BT(de Beer, Dahlmans & Smeets 1976, Geneste et al. 1980 and Van Parijs et al. 1986), DBT (Ho & Sobel 1991, Kim et al. 2005 and Mijoin et al. 2001), and their derivatives have been widely studied. The HDS of sulfur molecules generally proceeds by two pathways: a hydrogenation (HYD) pathway involving aromatic ring hydrogenation, followed by C-S bond cleavage, and a hydrogenolysis pathway via direct C-S bond cleavage without aromatic ring hydrogenation, which is also called the direct desulfurization (DDS) pathway. Combining DFT and temperature programmed desorption (TPD), we have recenlty proved that in the H_2 environment, thiophene desulfurization is clearly through HYD, rather than DDS. Further details of the evidence can be found in our to-be published work (Liu et al. 2016).

A more realistic insight into the HDS reaction of thiophene was provided by Kumar & Seminario (2015) used a multi-scale modelling approach. As Figure 8 shows, a simulation box is established containing a mixture of thiophene, gaseous hydrogen and some other aromatic molecules (denoted as "hydrocarbon oil") between two layers of MoS_2 catalyst. Molecular dynamics (MD) simulations have identified the catalyst edges as the active sites where thiophene and H_2 molecules aggregate.

Figure 8. Unit cell box for periodic MD simulations in three dimensions. The unit cell contains 217 n-C15, 369 n-nonylhexane, 130 n-nonylbenzene, 152 naphthalene, and 217 thiophene, with 108 H_2 molecules sandwiched between two parallel MoS_2 slabs; each MoS_2 slab contains 3456 Mo and 6912 S atoms, making a total of 57515 atoms for the full unit cell.
Source: Reproduced with permission from Kumar & Seminario (2015). Copyright of American Chemical Society

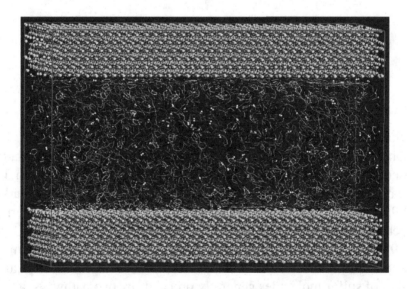

2. Mo₂C NANOCATALYSTS

It has been found since 1970s that transition metal carbides (TMC) have comparable activity to group VIII noble metals in many surface catalytic reactions. They are proved to be good catalysts in a series of heterogeneous catalytic reactions like hydrogenation (Shahrzad Jooya Ardakani et al. 2007, Ardakani & Smith, 2011, Liu et al. 2011 and Lee et al. 1991), n-butane hydrogenolysis (Leclercq et al. 1989), ammonia synthesis (Oyama 1992), hydrodehalogenation (Oxley et al. 2004) and aromatization (Róbert et al. 2007; Barthos & Solymosi 2005 and Solymosi & Barthos 2005). In addition, they are very suitable as catalysts because of their excellent physical properties e.g. extreme hardness, high melting point, and high electric and thermal conductivity. It is thus believed that they have the potential to become cheap substitutes for the noble metals in heterogeneous catalysis. In particular, molybdenum carbide has been an active catalyst in hydrodesulfurization (McCrea et al. 1997), hydrodenitrogenation (Schlatter et al. 1988) and hydrogenation (Marquez Alvarez et al. 1997).

2.1 Bulk Mo₂C

Molybdenum carbides are crystalline compounds of metallic characteristics that are formed by the incorporation of carbon into the lattice of molybdenum. The crystal structure of molybdenum carbide is usually arranged in a close-packed structure, with carbon occupying the interstitial octahedral sites. It has two stable forms, hcp (β-Mo$_2$C, hexagonal) and fcc (α-Mo$_2$C, orthorhombic). The difference in hardness and melting point of the two phases can be attributed to the stronger hybridization effects between Mo-3d and C-2p states, in other words, stronger bonding between Mo and C in α-Mo$_2$C. Although the stoichiometry of Mo and C is formally 2:1, the material is always carbon deficit, sometimes written as Mo$_2$C$_{1-x}$. It is estimated based on DFT calculations of the Helmholtz free energy that at 650 K, 9.6% of molybdenum vacancies and 10.2% of carbon vacancies exist (de Oliveira et al. 2014).

The exposed Mo$_2$C surfaces can have either Mo or C termination. DFT studies show that for different surfaces, the most stable terminations are different. For example, for (001) and (100) surface of β-Mo$_2$C, the pure C terminations are more stable than pure Mo terminations, while for (010) and (111) surface, the opposite is true. Mixed Mo/C termination is the most stable one for the (011) and (101) surfaces (Shi et al. 2009).

2.2 Mo$_2$C Nanoparticles

For catalytic applications, ultra-dispersed nanoparticles with high surface area are preferred. Experimentally, molybdenum carbide nanoparticles of 2-3 nm in size have a stoichiometry of Mo$_2$C$_{0.97}$ (Hyeon et al. 1996). The structures and energy properties of small molybdenum carbide clusters (Mo$_2$C)$_n$, n=1-10 was predicted by Born-Oppenheimer molecular dynamics (BOMD) simulations(Cruz-Olvera & Calaminici 2016). When the carbon atoms are inserted with stoichiometry of Mo:C = 2:1, the ground state structure of (Mo$_2$C)$_n$, n=1-4 largely maintains the molybdenum framework of the bare clusters(Elliott et al. 2009). Compared to the bulk crystalline structure, the (Mo$_2$C)$_n$ with more than six atoms shows clear packing effects. The spin density of the clusters is almost completely localized on the Mo atoms.

Liu et al (2015b) used density functional tight-binding (DFTB) method to study the equilibrium shape of Mo$_2$C nanoparticles of 1.2 nm, 1.9 nm, and 2.3 nm. The structures obtained are mostly spherical-like in topology. The nanoparticles of all three sizes shows clear metallic nature with very small HOMO-LUMO gap (0.0012 eV to 0.0047 eV).

For larger nanoparticles, the equilibrium structure of β-Mo$_2$C nanoparticles at various working conditions were predicted based on ab initio atomistic thermodynamic calculations and Wulff construction (Wang et al. 2011) (Figure 9). At low temperature, the β-Mo$_2$C nanoparticles prefer to expose the (001) surface, while at high temperature, the (101) surface becomes dominant, which is in agreement with the High-resolution transmission electron microscopy (HRTEM) experiment (Nagai et al. 2006). Upon Alkali-promotion, (K and Rb), a surface reconstruction happens. On alkali-promoted surface, the (011) surface has the lowest surface energy, and is the one that is mostly exposed(Han et al. 2011). Using the same method, α-Mo$_2$C was predicted to most likely expose the (111) surface, and the equilibrium shape of the nanoparticle is sensitive to hydrogen adsorption (Wang et al. 2016).

2.3 The Adsorption of Small Molecules on Mo$_2$C Surfaces

Ren et al. (2005, 2006 and 2007) systematically studied the adsorption of oxygen atom and CO, CO$_2$, H and CH$_x$ (x=0-3) as well as NO and NO$_2$ on Mo- and C-terminated α-Mo$_2$C (0001) surfaces. On Mo-terminated surface, both H and the O atoms and the CH$_x$ species prefer the three-fold hollow site with a second layer carbon atom. On C-terminated surface, O atom prefer three-fold hollow sites on three Mo atoms without a C atom on top. As the

Figure 9. The Wulff shapes and proportion of surfaces areas of hexagonal Mo_2C under different conditions: (a) for the shape at 0, 600, and 1000 K under CH_4/H_2 = 1/4 at 1 atm, and (b) for the shape at 0, 600, and 1000 K under CO/CO_2 = 2/1 at 1 atm.
Source: Reproduced with permission from Wang et al. (2011). Copyright of American Chemical Society.

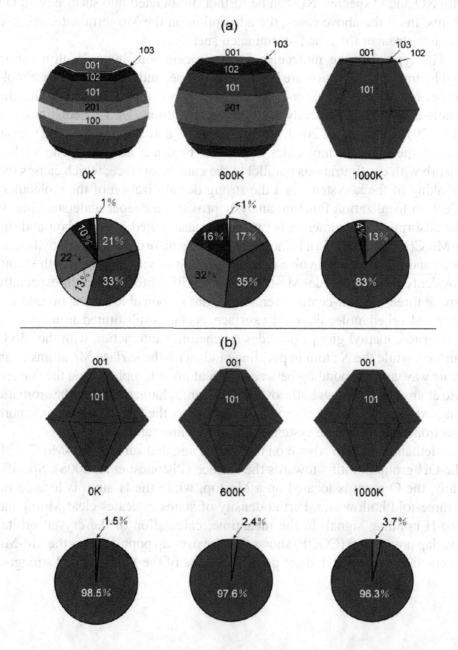

number of H atoms in CH_x (x=0-3) increase, the binding strength decreases. For CO, the most stable configuration is that also a three-fold hollow site on Mo-terminated surface, but a C top site on the C-terminated surface. Upon adsorption, CO_2 dissociate into CO and O species, and NO_2 decomposes into NO and O species. NO can be further dissociated into surface N and O atoms. In all the above cases, the adsorption on the Mo-terminated surface is much stronger than the C-terminated surface.

The typical probe molecules for hydrogenation, hydrodesulfurization and hydrodenitrogenation are benzene, pyridine, and thiophene. These molecules are all very stable π-systems that can adsorb horizontally with the whole π-system or vertically with only a few atoms. DFT calculations(Liu et al. 2015a, Liu et al. 2013, Ren et al. 2007, and Zhou et al. 2012) show that all the aromatic molecules - pyridine, benzene, and thiophene tend to adsorb with configurations parallel to the catalyst surface, which causes the breaking of the π-systems and the strong destabilization of the molecules. Electron localization function analysis provides a deeper understanding of the adsorption phenomenoa between the unsaturated hydrocarbon and the α-Mo_2C(Liu et al. 2015a). It shows that the chemisorption between hydrocarbons and the MCNPs involve electron sharing of various types with strong covalent character. Besides Mo-C σ-bonds, multi-center interaction, especially strong three- or four-center interactions, are responsible for the orientation of the adsorbed molecules on the surface. As far as substituted aromatics are concerned, methyl groups provides no chemical interaction with the Mo_2C surface, while the N atom in pyridine binds with the surface Mo atoms in an ionic way, and the bonding between the S atom in thiophene and the surface Mo atoms clearly shows both covalent and ionic character. When the aromatic ring extends, say from benzene to naphthalene, the C-C bonds retains more electrons and the whole system retains more aromatic nature.

Methanol tends to adsorb on the Mo-terminated surface of β-Mo_2C with the OH group pointing towards the surface (Pistonesi et al. 2008). Specifically, the O atom is located on a Mo top, while the H atom is located on a three-fold hollow site. Partial density of states indicates clear Mo-O and Mo-H bonding signal. In the meantime, calculation of the crystal orbital overlap population (COOP) shows that the overlap population of the Mo-Mo interaction decreases, indicating the decrease of the Mo-Mo bond strength.

2.4 The Mechanism of Mo$_2$C Nanoparticles as Hydrogenation Catalysts from Theoretical Perspectives

As a hydrogenation catalyst, the working mechanism of Mo$_2$C nanoparticles in the hydrogenation of benzene, a model aromatic molecule, was studied systematically by Liu and Salahub et al. (Liu et al. 2015) using theoretical modelling. Early works with cluster (Liu et al. 2013) and periodic (Zhou et al. 2012) DFT methods suggest that benzene hydrogenation on Mo$_2$C happens through the Langmuir-Hinshelwood mechanism. The benzene molecule adsorbs horizontally on the three-fold site of the Mo-terminated surface, causing the broken of the π system and the destabilization of the molecule.

Figure 10. Optimized structures of the consecutive steps of benzene hydrogenation on a 3-fold site of the 1.2 nm MCNP.
Source: Reproduced with permission from Liu et al. (2015). Copyright of American Chemical Society

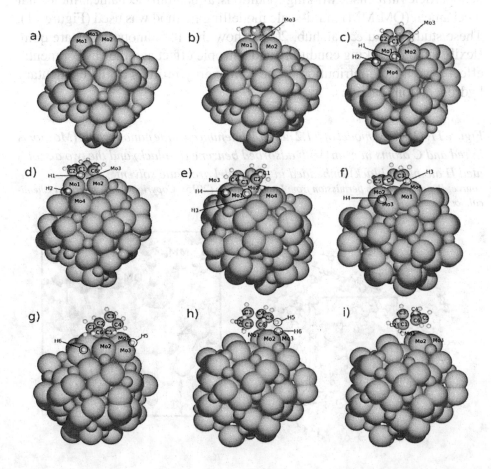

During the hydrogenation, the six-member ring tilt up gradually, and the H atoms are supplied constantly on the surface from the dissociative adsorption of H_2. The 1,2-C_6H_8, 1,2,3,4-C_6H_{10} species were identified as the most possible reaction intermediates.

A tight-binding quantum chemical molecular dynamics (TB-QCMD) method was used to track the physical motion of the atoms in the reaction processes (Ahmed et al. 2015). The pre-adsorbed benzene (C_6H_6) molecule was transformed to cyclohexadiene (C_6H_8), and then cyclohexane (C_6H_{12}), followed by a desorption, which confirmed the proposed Langmuir-Hinshelwood mechanism and the intermediate species in the previous works (Liu et al. 2013 and Zhou et al., 2012). The relevant reaction barriers of benzene hydrogenation on a 1.2 nm Mo_2C nanoparticle were calculated by density functional tight-binding (DFTB) method (Figure 10), and the sixth step (the last hydrogenation step) was found to be the one with highest barrier.

To provide insights into the mechanism of benzene hydrogenation on Mo_2C nanoparticles in realistic working conditions, a quantum mechanical/molecular mechanical (QM/MM) multi-scale modelling method was used (Figure 11). These studies (Liu & Salahub, 2015) show that the nanoparticles are quite flexible in the working condition, and entropic effect and the environmental effect have large contribution to the free energy profiles of the elementary hydrogenation reactions.

Figure 11. QM/MM model of a 1.2 nm molybdenum carbide nanoparticle (Mo atoms in red and C atoms in cyan) with adsorbed benzene (in black) and the two dissociated H atoms (in black) embedded in the model aromatic solvent.
Source: Reproduced with permission from Liu & Salahub (2015). Copyright 2017 American Chemical Society.

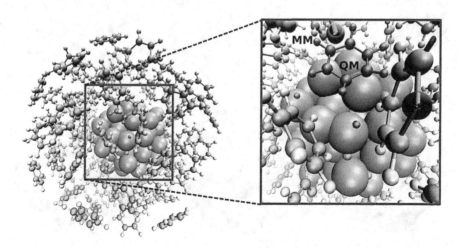

3. MoN₂ 2D CATALYTIC MATERIALS

Molybdenum nitride is another type of effective hydro-treating catalysts (Choi et al. 1994 and Colling et al. 1996). However, most of the molybdenum nitride catalyst used are nitrogen-deficit Mo_2N. Inspired by the high HDS activity of 3R-MoS_2, Wang et al. (2015) recently prepared a nitrogen-rich layered MoN_2 catalyst denoted as 3R–MoN_2, which can be a promising substitute of MoS_2 for HDS reactions. Soon theoretical studies discovered the reason for the existing of the 2D structure by the magnetism of the material (Wu et al. 2015). As Figure 12 shows, each monolayer of MoN_2 is ferromagnetic. However, the layers are stacked in such a way in 3R-MoN_2 that the magnetism of the layers cancel with one another, and the bulk has an antiferromagnetic ground state. Compared to other magnetic materials where the spin moments is often from the metal ions, 3R-MoN_2 is unique in that the spin moments mainly come from the p_z orbitals of the nitrogen ions. It is the symmetry of nitrogen p_z orbitals that make the intra-layer bonding ferromagnetic, and the inter-layer bonding antiferromagnetic. The net magnetic moments per N and Mo ions are calculated as 0.42 µB and 0.14 µB, respectively.

Although with novel structures and promising applications, this above 3R-MoN_2 is not the global minimum of MoN_2. Using USPEX structure search method, it was predicted (Yu et al. 2016) that the ground state of MoN_2 is a pernitride structure with space group $P6_3$/mmc which transforms to a P4/mbm phase above 82 GPa. Wang et al (2015) also found the existence of soft

Figure 12. (a) Spin density of ground states for bulk 3R–MoN_2, where green and yellow colors represent the spin-up and spin-down electrons, respectively. (b) Spin-polarized band structures of bulk 3R–MoN_2 in the antiferromagnetic ground state. (c) Spin-polarized band structures of bulk 3R–MoN_2 in the ferromagnetic state. Red and blue lines represent the spin-up and spin-down states, respectively.
Source: Reproduced with permission from Wu et al. (2015). Copyright of American Chemical Society.

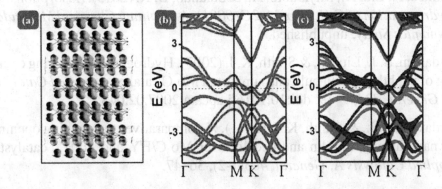

modes in 2D MoN_2. These soft modes, however, can be easily eliminated by surface hydrogenation, forming MoN_2H_2 sheets with good dynamical, thermal and mechanical stability. The applicability of 2D MoN_2 as catalysts for hydro-treating reactions is therefore valid. Calculation (Zhang et al. 2016) shows that the exfoliation energy of MoN_2 is 0.17 eV per unit cell, which is very close to that of MoS_2 (0.16 eV per unit cell). This suggests that 2D MoN_2 can be prepared experimentally using similar method to that of 2D MoS_2. In terms of electronic structure, the isolated monolayer MoN_2 is also metallic, and most of the spin-polarized states are contributed by the p_z orbitals of N ions (Wu et al. 2015). Besides the potential in hydro-treating catalysis, because of the metallic nature of the MoN_2, it can also be used as electrode material.

MoN_2 is a type of newly discovered 2D material. Early experiments and theoretical calculations have shown that it has the potential to be a promising 2D catalytic material with unique magnetic character. Its applications as a low-dimensional catalyst in hydro-treating and other reactions will be discovered in the future. The theory reveal to the 0D molybdenum carbide (Mo_2C) nanoparticle, the 2D molybdenum sulfide (MoS_2) as well as molybdenum nitride (MoN_2) have help us understand their structures, electronic properties and the mechanism of catalytic reactions. We are now close to a stage when high quality catalysts can be predicted based on theoretical structure-activity relationship, with the electronic structure as a bridge. Moreover, the structures and the mechanism of the real size highly-dispersed catalyst particles in the working conditions can be understood. Tailing molybdenum-based catalysts based on theoretical insights will enable more accurate and efficient development of these catalysts for HDS and many other industrial applications.

REFERENCES

Ahmed, F., Liu, X., Miyamoto, A., & Salahub, D. R. (2015). *Adsorption and Hydrogenation of C_6H_6 on α-Mo2C (0001): A Quantum Chemical Molecular Dynamics Study*. unpublished.

Ardakani, S. J., Liu, X., & Smith, K. J. (2007). Hydrogenation and ring opening of naphthalene on bulk and supported Mo_2C catalysts. *Applied Catalysis A, General, 324*, 9–19. doi:10.1016/j.apcata.2007.02.048

Ardakani, S. J., & Smith, K. J. (2011). A comparative study of ring opening of naphthalene, tetralin and decalin over Mo_2C/HY and Pd/HY catalysts. *Applied Catalysis A, General, 403*(1-2), 36–47.

Bar-Sadan, M., Enyashin, A. N., Gemming, S., Popovitz-Biro, R., Hong, S. Y., Prior, Y., & Seifert, G. et al. (2006). Structure and Stability of Molybdenum Sulfide Fullerenes. *The Journal of Physical Chemistry B, 110*(50), 25399–25410. doi:10.1021/jp0644560 PMID:17165987

Barthos, R., Bánsági, T., Süli Zakar, T., & Solymosi, F. (2007). Aromatization of methanol and methylation of benzene over Mo_2C/ZSM-5 catalysts. *Journal of Catalysis, 247*(2), 368–378. doi:10.1016/j.jcat.2007.02.017

Barthos, R., & Solymosi, F. (2005). Aromatization of n-heptane on Mo_2C-containing catalysts. *Journal of Catalysis, 235*(1), 60–68. doi:10.1016/j.jcat.2005.07.004

Borges, I. Jr, Silva, A. M., Aguiar, A. P., Borges, L. E. P., Santos, J. C. A., & Dias, M. H. C. (2007). Density functional theory molecular simulation of thiophene adsorption on MoS_2 including microwave effects. *Journal of Molecular Structure THEOCHEM, 822*(1–3), 80–88. doi:10.1016/j.theochem.2007.07.020

Chary, K. V. R., Ramakrishna, H., Rama Rao, K. S., Murali Dhar, G., & Kanta Rao, P. (1991). Hydrodesulfurization on MoS_2/MgO. *Catalysis Letters, 10*(1), 27–33. doi:10.1007/BF00764733

Chhowalla, M., Shin, H. S., Eda, G., Li, L.-J., Loh, K. P., & Zhang, H. (2013). The chemistry of two-dimensional layered transition metal dichalcogenide nanosheets. *Nature Chemistry, 5*(4), 263–275. doi:10.1038/nchem.1589 PMID:23511414

Choi, J.-G., Curl, R. L., & Thompson, L. T. (1994). Molybdenum nitride catalysts. *Journal of Catalysis, 146*(1), 218–227. doi:10.1016/0021-9517(94)90025-6

Coleman, J. N., Lotya, M., ONeill, A., Bergin, S. D., King, P. J., Khan, U., & Nicolosi, V. et al. (2011). Two-Dimensional Nanosheets Produced by Liquid Exfoliation of Layered Materials. *Science, 331*(6017), 568–571. doi:10.1126/science.1194975 PMID:21292974

Colling, C. W., Choi, J.-G., & Thompson, L. T. (1996). Molybdenum Nitride Catalysts. *Journal of Catalysis, 160*(1), 35–42. doi:10.1006/jcat.1996.0121

Cristol, S., Paul, J.-F., Payen, E., Bougeard, D., Hutschka, F., & Clémendot, S. (2004). DBT derivatives adsorption over molybdenum sulfide catalysts: A theoretical study. *Journal of Catalysis, 224*(1), 138–147. doi:10.1016/j.jcat.2004.02.008

Cristol, S., Paul, J.-F., Schovsbo, C., Veilly, E., & Payen, E. (2006). DFT study of thiophene adsorption on molybdenum sulfide. *Journal of Catalysis, 239*(1), 145–153. doi:10.1016/j.jcat.2006.01.015

Cruz-Olvera, D., & Calaminici, P. (2016). Investigation of structures and energy properties of molybdenum carbide clusters: Insight from theory. *Computational and Theoretical Chemistry, 1078*, 55–64. doi:10.1016/j.comptc.2015.12.019

de Beer, V. H. J., Dahlmans, J. G. J., & Smeets, J. G. M. (1976). Hydrodesulfurization and hydrogenation properties of promoted MoS_2 and WS_2 catalysts under medium pressure conditions. *Journal of Catalysis, 42*(3), 467–470. doi:10.1016/0021-9517(76)90122-6

de Oliveira, C., Salahub, D. R., de Abreu, H. A., & Duarte, H. A. (2014). Native Defects in α-Mo2C: Insights from First-Principles Calculations. *The Journal of Physical Chemistry C, 118*(44), 25517–25524. doi:10.1021/jp507947b

Devanneaux, J., & Maurin, J. (1981). Hydrogenolysis and hydrogenation of thiophenic compounds on a $Co-Mo/Al_2O_3$ catalyst. *Journal of Catalysis, 69*(1), 202–205. doi:10.1016/0021-9517(81)90142-1

Elliott, J. A., Shibuta, Y., & Wales, D. J. (2009). Global minima of transition metal clusters described by Finnis–Sinclair potentials: A comparison with semi-empirical molecular orbital theory. *Philosophical Magazine, 89*(34-36), 3311–3332. doi:10.1080/14786430903270668

Fleischauer, P. D., & Lince, J. R. (1999). A comparison of oxidation and oxygen substitution in MoS2 solid film lubricants. *Tribology International, 32*(11), 627–636. doi:10.1016/S0301-679X(99)00088-2

Galea, N. M., Kadantsev, E. S., & Ziegler, T. (2009). Modeling Hydrogen Sulfide Adsorption on Mo-Edge MoS2 Surfaces under Solid Oxide Fuel Cell Conditions. *The Journal of Physical Chemistry C, 113*(1), 193–203. doi:10.1021/jp711202q

Geneste, P., Amblard, P., Bonnet, M., & Graffin, P. (1980). Hydrodesulfurization of oxidized sulfur compounds in benzothiophene, methylbenzothiophene, and dibenzothiophene series over $CoO-MoO_3-Al_2O_3$ catalyst. *Journal of Catalysis, 61*(1), 115–127. doi:10.1016/0021-9517(80)90346-2

Gong, Y., Lin, J., Wang, X., Shi, G., Lei, S., Lin, Z., & Ajayan, P. M. et al. (2014). Vertical and in-plane heterostructures from WS_2/MoS_2 monolayers. *Nature Materials, 13*(12), 1135–1142. doi:10.1038/nmat4091 PMID:25262094

Han, J. W., Li, L., & Sholl, D. S. (2011). Density Functional Theory Study of H and CO Adsorption on Alkali-Promoted Mo_2C Surfaces. *The Journal of Physical Chemistry C, 115*(14), 6870–6876. doi:10.1021/jp200950a

Hargreaves, A. E., & Ross, J. R. H. (1979). An investigation of the mechanism of the hydrodesulfurization of thiophene over sulfided $CoMoAl_2O_3$ catalysts. *Journal of Catalysis, 56*(3), 363–376. doi:10.1016/0021-9517(79)90129-5

Helveg, S., Lauritsen, J. V., Lægsgaard, E., Stensgaard, I., Nørskov, J. K., Clausen, B. S., & Besenbacher, F. et al. (2000). Atomic-Scale Structure of Single-Layer MoS_2 Nanoclusters. *Physical Review Letters, 84*(5), 951–954. doi:10.1103/PhysRevLett.84.951 PMID:11017413

Hensen, E. J. M., Vissenberg, M. J., de Beer, V. H. J., van Veen, J. A. R., & van Santen, R. A. (1996). Kinetics and Mechanism of Thiophene Hydrodesulfurization over Carbon-Supported Transition Metal Sulfides. *Journal of Catalysis, 163*(2), 429–435. doi:10.1006/jcat.1996.0344

Ho, T. C., & Sobel, J. E. (1991). Kinetics of dibenzothiophene hydrodesulfurization. *Journal of Catalysis, 128*(2), 581–584. doi:10.1016/0021-9517(91)90316-V

Hyeon, T., Fang, M., & Suslick, K. S. (1996). Nanostructured Molybdenum Carbide: Sonochemical Synthesis and Catalytic Properties. *Journal of the American Chemical Society, 118*(23), 5492–5493. doi:10.1021/ja9538187

Kc, S., Longo, R. C., Wallace, R. M., & Cho, K.KC. (2015). Surface oxidation energetics and kinetics on MoS_2 monolayer. *Journal of Applied Physics, 117*(13), 135301. doi:10.1063/1.4916536

Kim, J. H., Ma, X., Song, C., Lee, Y.-K., & Oyama, S. T. (2005). Kinetics of Two Pathways for 4,6-Dimethyldibenzothiophene Hydrodesulfurization over NiMo, CoMo Sulfide, and Nickel Phosphide Catalysts. *Energy & Fuels, 19*(2), 353–364. doi:10.1021/ef049804g

Kobayashi, K., & Yamauchi, J. (1995). Electronic structure and scanning-tunneling-microscopy image of molybdenum dichalcogenide surfaces. *Physical Review B: Condensed Matter and Materials Physics, 51*(23), 17085–17095. doi:10.1103/PhysRevB.51.17085 PMID:9978722

Kormányos, A., Zólyomi, V., Drummond, N. D., & Burkard, G. (2014). Spin-Orbit Coupling, Quantum Dots, and Qubits in Monolayer Transition Metal Dichalcogenides. *Physical Review X, 4*(1), 011034. doi:10.1103/PhysRevX.4.011034

Kumar, N., & Seminario, J. M. (2015). Computational Chemistry Analysis of Hydrodesulfurization Reactions Catalyzed by Molybdenum Disulfide Nanoparticles. *The Journal of Physical Chemistry C, 119*(52), 29157–29170. doi:10.1021/acs.jpcc.5b09712

Lauritsen, J. V., Bollinger, M. V., Laegsgaard, E., Jacobsen, K. W., Norskov, J. K., Clausen, B. S., & Besenbacher, F. et al. (2004). Atomic-scale insight into structure and morphology changes of MoS_2 nanoclusters in hydrotreating catalysts. *Journal of Catalysis, 221*(2), 510–522. doi:10.1016/j.jcat.2003.09.015

Leclercq, L., Provost, M., Pastor, H., & Leclercq, G. (1989). Catalytic properties of transition metal carbides: II. Activity of bulk mixed carbides of molybdenum and tungsten in hydrocarbon conversion. *Journal of Catalysis, 117*(2), 384–395. doi:10.1016/0021-9517(89)90349-7

Lee, J. S., Yeom, M. H., Park, K. Y., Nam, I.-S., Chung, J. S., Kim, Y. G., & Moon, S. H. (1991). Preparation and benzene hydrogenation activity of supported molybdenum carbide catalysts. *Journal of Catalysis, 128*(1), 126–136. doi:10.1016/0021-9517(91)90072-C

Liang, T., Sawyer, W. G., Perry, S. S., Sinnott, S. B., & Phillpot, S. R. (2011). Energetics of Oxidation in MoS_2 Nanoparticles by Density Functional Theory. *The Journal of Physical Chemistry C, 115*(21), 10606–10616. doi:10.1021/jp110562n

Liu, A. C., & Friend, C. M. (1991). Evidence for facile and selective desulfurization: The reactions of 2,5-dihydrothiophene on molybdenum(110). *Journal of the American Chemical Society, 113*(3), 820–826. doi:10.1021/ja00003a014

Liu, X., Cao, D., Peng, Q., Ge, H., Ramos, M., & Wen, X. (2016). *Insights into Structure and Energy of $Mo_{27}S_xO_y$ cluster*. In preparation.

Liu, X., Feng, J., Ge, H., Cao, D., Yang, T., Li, B.,... Wen, X. (2016). *DFT and TPD study of Sulfur Containing Compounds Adsorption on MoS_2: Competition, Hydrogenation and Steric effect*. In preparation.

Liu, X., & Salahub, D. R. (2015a). Application of topological analysis of the electron localization function to the complexes of molybdenum carbide nanoparticles with unsaturated hydrocarbons. *Canadian Journal of Chemistry, 94*(4), 282–292. doi:10.1139/cjc-2015-0075

Liu, X., & Salahub, D. R. (2015b). Molybdenum Carbide Nanocatalysts at Work in the in Situ Environment: A Density Functional Tight-Binding and Quantum Mechanical/Molecular Mechanical Study. *Journal of the American Chemical Society*, *137*(12), 4249–4259. doi:10.1021/jacs.5b01494 PMID:25774905

Liu, X., Tkalych, A., Zhou, B., Köster, A. M., & Salahub, D. R. (2013). Adsorption of Hexacyclic C_6H_6, C_6H_8, C_6H_{10}, and C_6H_{12} on a Mo-Terminated α-Mo_2C (0001) Surface. *The Journal of Physical Chemistry C*, *117*(14), 7069–7080. doi:10.1021/jp312204u

Liu, X., Zhou, B., Ahmed, F., Tkalych, A., Miyamoto, A., & Salahub, D. (2015). Multiscale Modelling of In Situ Oil Sands Upgrading with Molybdenum Carbide Nanoparticles. In J.-L. Rivail, M. Ruiz-Lopez, & X. Assfeld (Eds.), *Quantum Modeling of Complex Molecular Systems* (Vol. 21, pp. 415–445). Springer International Publishing. doi:10.1007/978-3-319-21626-3_16

Liu, X. B., Ardakani, S. J., & Smith, K. J. (2011). The effect of Mg and K addition to a Mo_2C/HY catalyst for the hydrogenation and ring opening of naphthalene. *Catalysis Communications*, *12*(6), 454–458. doi:10.1016/j.catcom.2010.10.025

Lu, P., Wu, X., Guo, W., & Zeng, X. C. (2012). Strain-dependent electronic and magnetic properties of MoS_2 monolayer, bilayer, nanoribbons and nanotubes. *Physical Chemistry Chemical Physics*, *14*(37), 13035–13040. doi:10.1039/c2cp42181j PMID:22911017

Mak, K. F., Lee, C., Hone, J., Shan, J., & Heinz, T. F. (2010). Atomically Thin MoS_2: A New Direct-Gap Semiconductor. *Physical Review Letters*, *105*(13), 136805. doi:10.1103/PhysRevLett.105.136805 PMID:21230799

Markel, E. J., Schrader, G. L., Sauer, N. N., & Angelici, R. J. (1989). Thiophene, 2,3- and 2,5-dihydrothiophene, and tetrahydrothiophene hydrodesulfurization on Mo and Re γ-Al_2O_3 catalysts. *Journal of Catalysis*, *116*(1), 11–22. doi:10.1016/0021-9517(89)90071-7

MarquezAlvarez, C., Claridge, J. B., York, A. P. E., Sloan, J., & Green, M. L. H. (1997). *Benzene hydrogenation over transition metal carbides* (Vol. 106). Academic Press.

McCrea, K. R., Logan, J. W., Tarbuck, T. L., Heiser, J. L., & Bussell, M. E. (1997). Thiophene Hydrodesulfurization over Alumina-Supported Molybdenum Carbide and Nitride Catalysts: Effect of Mo Loading and Phase. *Journal of Catalysis*, *171*(1), 255–267. doi:10.1006/jcat.1997.1805

Mijoin, J., Pérot, G., Bataille, F., Lemberton, J. L., Breysse, M., & Kasztelan, S. (2001). Mechanistic considerations on the involvement of dihydrointermediates in the hydrodesulfurization of dibenzothiophene-type compounds over molybdenum sulfide catalysts. *Catalysis Letters, 71*(3-4), 139–145. doi:10.1023/A:1009055205076

Moses, P. G., Mortensen, J. J., Lundqvist, B. I., & Nørskov, J. K. (2009). Density functional study of the adsorption and van der Waals binding of aromatic and conjugated compounds on the basal plane of MoS_2. *The Journal of Chemical Physics, 130*(10), 104709. doi:10.1063/1.3086040 PMID:19292551

Nagai, M., Zahidul, A. M., & Matsuda, K. (2006). Nano-structured nickel–molybdenum carbide catalyst for low-temperature water-gas shift reaction. *Applied Catalysis A, General, 313*(2), 137–145. doi:10.1016/j.apcata.2006.07.006

Orita, H., Uchida, K., & Itoh, N. (2003a). Ab initio density functional study of the structural and electronic properties of an MoS_2 catalyst model: A real size $Mo_{27}S_{54}$ cluster. *Journal of Molecular Catalysis A Chemical, 195*(1–2), 173–180. doi:10.1016/S1381-1169(02)00528-9

Orita, H., Uchida, K., & Itoh, N. (2003b). Adsorption of thiophene on an MoS_2 cluster model catalyst: Ab initio density functional study. *Journal of Molecular Catalysis A Chemical, 193*(1–2), 197–205. doi:10.1016/S1381-1169(02)00467-3

Oxley, J. D., Mdleleni, M. M., & Suslick, K. S. (2004). Hydrodehalogenation with sonochemically prepared Mo_2C and W_2C. *Catalysis Today, 88*(3–4), 139–151. doi:10.1016/j.cattod.2003.11.010

Oyama, S. T. (1992). Preparation and catalytic properties of transition metal carbides and nitrides. *Catalysis Today, 15*(2), 179–200. doi:10.1016/0920-5861(92)80175-M

Pistonesi, C., Juan, A., Farkas, A. P., & Solymosi, F. (2008). DFT study of methanol adsorption and dissociation on β-Mo_2C(001). *Surface Science, 602*(13), 2206–2211. doi:10.1016/j.susc.2008.04.039

Rangarajan, S., & Mavrikakis, M. (2016). DFT Insights into the Competitive Adsorption of Sulfur- and Nitrogen-Containing Compounds and Hydrocarbons on Co-Promoted Molybdenum Sulfide Catalysts. *ACS Catalysis, 6*(5), 2904–2917. doi:10.1021/acscatal.6b00058

Ren, J., Huo, C.-F., Wang, J., Li, Y.-W., & Jiao, H. (2005). Surface structure and energetics of oxygen and CO adsorption on α-$Mo_2C(0001)$. *Surface Science*, *596*(1–3), 212–221. doi:10.1016/j.susc.2005.09.018

Ren, J., Huo, C. F., Wang, J. G., Cao, Z., Li, Y. W., & Jiao, H. J. (2006). Density functional theory study into the adsorption of CO_2, H and CH_x (x=03) as well as C_2H_4 on alpha-$Mo_2C(0001)$. *Surface Science*, *600*(11), 2329–2337. doi:10.1016/j.susc.2006.03.027

Ren, J., Wang, J. G., Huo, C. F., Wen, X. D., Cao, Z., Yuan, S. P., & Jiao, H. J. et al. (2007). Adsorption of NO, NO2, pyridine and pyrrole on alpha-Mo2C(0001): A DFT study. *Surface Science*, *601*(6), 1599–1607. doi:10.1016/j.susc.2007.01.036

Roberts, J. T., & Friend, C. M. (1986). Model hydrodesulfurization reactions: Saturated tetrahydrothiophene and 1-butanethiol on molybdenum(110). *Journal of the American Chemical Society*, *108*(23), 7204–7210. doi:10.1021/ja00283a011

Schlatter, J. C., Oyama, S. T., Metcalfe, J. E., & Lambert, J. M. (1988). Catalytic behavior of selected transition metal carbides, nitrides, and borides in the hydrodenitrogenation of quinoline. *Industrial & Engineering Chemistry Research*, *27*(9), 1648–1653. doi:10.1021/ie00081a014

Schweiger, H., Raybaud, P., Kresse, G., & Toulhoat, H. (2002). Shape and edge sites modifications of MoS_2 catalytic nanoparticles induced by working conditions: A theoretical study. *Journal of Catalysis*, *207*(1), 76–87. doi:10.1006/jcat.2002.3508

Sen, H. S., Sahin, H., Peeters, F. M., & Durgun, E. (2014). Monolayers of MoS_2 as an oxidation protective nanocoating material. *Journal of Applied Physics*, *116*(8), 083508. doi:10.1063/1.4893790

Shi, X.-R., Wang, S.-G., Wang, H., Deng, C.-M., Qin, Z., & Wang, J. (2009). Structure and stability of β-Mo_2C bulk and surfaces: A density functional theory study. *Surface Science*, *603*(6), 852–859. doi:10.1016/j.susc.2009.01.041

Solymosi, F., & Barthos, R. (2005). Aromatization of n-hexane on Mo_2C catalysts. *Catalysis Letters*, *101*(3-4), 235–239. doi:10.1007/s10562-005-4898-y

Stefanov, M., Enyashin, A. N., Heine, T., & Seifert, G. (2008). Nanolubrication: How Do MoS_2-Based Nanostructures Lubricate? *The Journal of Physical Chemistry C*, *112*(46), 17764–17767. doi:10.1021/jp808204n

Sullivan, D. L., & Ekerdt, J. G. (1998). Mechanisms of Thiophene Hydrodesulfurization on Model Molybdenum Catalysts. *Journal of Catalysis, 178*(1), 226–233. doi:10.1006/jcat.1998.2162

Travert, A., Dujardin, C., Mauge, F., Veilly, E., Cristol, S., Paul, J. F., & Payen, E. (2006). CO adsorption on CoMo and NiMo sulfide catalysts: A combined IR and DFT. *The Journal of Physical Chemistry B, 110*(3), 1261–1270. doi:10.1021/jp0536549 PMID:16471673

Van Parijs, I. A., Hosten, L. H., & Froment, G. F. (1986). Kinetics of the hydrodesulfurization on a cobalt-molybdenum/.gamma.-alumina catalyst. 2. Kinetics of the hydrogenolysis of benzothiophene. *Industrial & Engineering Chemistry Product Research and Development, 25*(3), 437–443. doi:10.1021/i300023a012

Wang, S., Ge, H., Sun, S., Zhang, J., Liu, F., Wen, X., & Zhao, Y. et al. (2015). A New Molybdenum Nitride Catalyst with Rhombohedral MoS2 Structure for Hydrogenation Applications. *Journal of the American Chemical Society, 137*(14), 4815–4822. doi:10.1021/jacs.5b01446 PMID:25799018

Wang, T., Liu, X., Wang, S., Huo, C., Li, Y.-W., Wang, J., & Jiao, H. (2011). Stability of β-Mo_2C Facets from ab Initio Atomistic Thermodynamics. *The Journal of Physical Chemistry C, 115*(45), 22360–22368. doi:10.1021/jp205950x

Wang, T., Tian, X. X., Yang, Y., Li, Y. W., Wang, J. G., Beller, M., & Jiao, H. J. (2016). Surface morphology of orthorhombic Mo_2C catalyst and high coverage hydrogen adsorption. *Surface Science, 651*, 195–202. doi:10.1016/j.susc.2016.04.017

Wen, X.-D., Zeng, T., Li, Y.-W., Wang, J., & Jiao, H. (2005). Surface Structure and Stability of MoS_x Model Clusters. *The Journal of Physical Chemistry B, 109*(39), 18491–18499. doi:10.1021/jp051540r PMID:16853381

Wen, X. D., Zeng, T., Li, Y. W., Wang, J. G., & Jiao, H. J. (2005). Surface structure and stability of MoS_x model clusters. *The Journal of Physical Chemistry B, 109*(39), 18491–18499. doi:10.1021/jp051540r PMID:16853381

Wen, X.-D., Zeng, T., Teng, B.-T., Zhang, F.-Q., Li, Y.-W., Wang, J., & Jiao, H. (2006). Hydrogen adsorption on a $Mo_{27}S_{54}$ cluster: A density functional theory study. *Journal of Molecular Catalysis A Chemical, 249*(1–2), 191–200. doi:10.1016/j.molcata.2006.01.018

Wu, F., Huang, C., Wu, H., Lee, C., Deng, K., Kan, E., & Jena, P. (2015). Atomically Thin Transition-Metal Dinitrides: High-Temperature Ferromagnetism and Half-Metallicity. *Nano Letters, 15*(12), 8277–8281. doi:10.1021/acs.nanolett.5b03835 PMID:26575002

Xiao, S. L., Yu, W. Z., & Gao, S. P. (2016). Edge preference and band gap characters of MoS_2 and WS_2 nanoribbons. *Surface Science, 653*, 107–112. doi:10.1016/j.susc.2016.06.011

Yu, S., Huang, B., Jia, X., Zeng, Q., Oganov, A. R., Zhang, L., & Frapper, G. (2016). Exploring the Real Ground-State Structures of Molybdenum Dinitride. *The Journal of Physical Chemistry C, 120*(20), 11060–11067. doi:10.1021/acs.jpcc.6b00665

Yun, W. S., Han, S. W., Hong, S. C., Kim, I. G., & Lee, J. D. (2012). Thickness and strain effects on electronic structures of transition metal dichalcogenides: 2H-MX_2 semiconductors (M=Mo, W; X=S, Se, Te). *Physical Review B: Condensed Matter and Materials Physics, 85*(3), 033305. doi:10.1103/PhysRevB.85.033305

Zahid, F., Liu, L., Zhu, Y., Wang, J., & Guo, H. (2013). A generic tight-binding model for monolayer, bilayer and bulk MoS_2. *Aip Advances, 3*(5), 052111. doi:10.1063/1.4804936

Zhang, X., Yu, Z., Wang, S.-S., Guan, S., Yang, H. Y., Yao, Y., & Yang, S. A. (2016). Theoretical prediction of MoN_2 monolayer as a high capacity electrode material for metal ion batteries. *Journal of Materials Chemistry A, 4*(39), 15224–15231. doi:10.1039/C6TA07065E

Zhou, B., Liu, X., Cuervo, J., & Salahub, D. R. (2012). Density functional study of benzene adsorption on the α-Mo_2C(0001) surface. *Structural Chemistry, 23*(5), 1459–1466. doi:10.1007/s11224-012-0064-5

Zhou, W., Zou, X., Najmaei, S., Liu, Z., Shi, Y., Kong, J., & Idrobo, J.-C. et al. (2013). Intrinsic Structural Defects in Monolayer Molybdenum Disulfide. *Nano Letters, 13*(6), 2615–2622. doi:10.1021/nl4007479 PMID:23659662

APPENDIX: LIST OF ACRONYMS (ALPHABETICAL ORDER)

CUS: Coordinatively Unsaturated Sites
DBT: Dibenzothiophene
DDS: Direct Desulfurization
DFT: Density Functional Theory
DFTB: Density Functional Tight-Binding
DHT: Dihydrothiphene
DOS: Density of States
HRTEM: High-Resolution Transmission Electron Microscopy
HYD: Hydrogenation
HOMO: Highest Occupied Molecular Orbital
LDOS: Local Density Of States
LUMO: lowest unoccupied molecular orbital
MD: Molecular dynamics
QM/MM: Quantum Mechanical/Molecular Mechanical
TPD: Temperature Programmed Desorption
THT: Tetrahydrothiophene
TB-QCMD: Tight-Binding Quantum Chemical Molecular Dynamics
4,6-DMDBT: 4,6-Dimethyl-Substituted Dibenzothiophene

Chapter 5
Summary and Perspectives

The social concerning for the energy and environment have push the petroleum and chemical industry forward to cleaner production and greener processes. Enhancement of process effectivity and decrease of pollutant emission are dominated to large extent by the improvement of related catalysis system. Mo(W) based catalysts are playing critical important roles in energy conversion and chemical production. The fast advance of fundamental research have brought the design and engineering of these catalysts at the nano, molecular and even atomic level.

1. THE SUMMARY

We have seen increasing efforts in developing new synthetic strategies for the growth of 2D Mo(W) dichalcogenides. Thermal CVD process have been evidenced to be effective. And the improved MOCVD can grow high quality of 2D layer Mo(W) dichalcogenides. Obtained materials are shown to be active in water splitting to produce hydrogen and CO_2 electric reduction. However, these preparing techniques are now limited for initial proof-of-concept studies of model catalysts owing to only small quantity available. On the contrary,

DOI: 10.4018/978-1-5225-2274-4.ch005

Copyright ©2017, IGI Global. Copying or distributing in print or electronic forms without written permission of IGI Global is prohibited.

the solvent or surfactant assisted exfoliation and wet chemical methods can prepare 2D Mo(W) dichalcogenides in larger quantity.

Electrochemical intercalation can effectively shift the chemical potentials of 2D Mo(W) dichalcogenides materials to the optimized position for efficient catalysis. Sometimes the intercalation process introduces a phase transition of the host matrix, which change catalytic activity and selectivity. Heterostructures of 2D Mo(W) dichalcogenides have been evidenced multicatalytic functions and adjustable properties However, the synthetic integration of multiple Mo(W) dichalcogenides to create desired heterostructure or superlattice is an uneasy job. And it is difficult to integrate of these 2D layered materials with other materials or support without damaging their lattice structure or altering their intrinsic electronic properties. 2D Mo(W) dichalcogenides and their derivatives have been active in the water splitting and CO_2 electrochemical reduction.

The 2D Mo(W)S_2 based catalysts are extensively used in the hydrotreating process for removal of heteroatoms and enhancement of oil quality. Insight into structure, mechanism and reactivity of these catalysts have recently gained using the combination of novel experimental and theoretical techniques, such as STM, DFT and HAADF. It is suggested that the hydrogenation reactions may take place on the brim sites, whereas the sulfur removal can take place at both Mo and S edges. And the STM results reveal the detailed structures of Ni–Mo–S and Co–Mo–S active phase and DFT reveal the Co-Mo-C active structure. And the relation of morphology and the electronic structure with Co and Ni promoters have also been resolved.

It has long been debated about the promoting mechanism of Co and Ni ions. Although the Co(Ni)-Mo-S model is accepted by most researchers, the Remote Control mechanism is still referred in some studies. In fact, these two models all have their sayings. Which one plays more important role depends on the preparation method and components of hydrotreating catalysts. And competition or synergy may exist in real catalyst system.

Meeting the stringent specifications represent one of the major challenges for the petroleum refining industry. Dropping of sulfur content of diesel fuel to very low levels, (e.g. 10 ppm) requires the removal of refractory sulfur species such as 4,6-DMDBT from the diesel stream. This issue is exacerbated by the inhibiting effect of polyaromatics, nitrogen compounds and the formed H_2S gas. Hydrogenation of aromatic cycle and isomerization of substituted groups can decrease the steric hindrance of 4,6-DMDBT molecule and facilitates the desulfurization. And hydrodenitrogenation (HDN) is also preferred for HDS of 4,6-DMDBT derivatives.

Summary and Perspectives

To CoMo, NiMo and NiW catalysts aiming to diesel deep HDS, the improvement can be achieved by increasing loading of active metal (Mo, W, etc.); by adding one more transient metal (e.g. Ni to CoMo or Co to NiMo); and by incorporating a noble metal (Pt, Pd, Ru, etc.). However, the favored strategies are fine control of the active structures and the interaction with support. The catalytic properties can be improved by using different supports (carbon, TiO_2, TiO_2-Al_2O_3, HY, MCM-41, etc.) which adjust the acidity and the interaction with active phases. Usage of chelating agent or P additives are also effective.

Another problem faced by the refiners is deep HDS of gasoline without the apparent drop of octane value. Contrary to the diesel HDS, Mo(W)S_2 based catalysts for gasoline HDS should be effective desulfurization of thiophene derivatives with least hydrogenation to olefins (HYDO). And some isomerization activity is preferred which can restore the octane number in some extent. Improved FCC naphtha HDS has been achieved over NiW catalyst where W-based hybrid nanocrystals are supported and promoted with Ni. It is reported that Co increase both HDS activity and HDS/HYDO selectivity. Meanwhile K increase HDS/HYDO selectivity accompanying the decrease of HDS activity.

Molybdenum sulfide catalysts are active for higher alcohol synthesis from CO hydrogenation due to the high selectivity to linear alcohols, slow deactivation, and low sensitivity to CO_2. Addition of transition and alkali metal is beneficial for the selective synthesis of higher alcohols. For avoidance the sulfur loss on catalyst, H_2S is usually added in feed which however brings in sulfur in product.

Apart from the Mo(W) dichalcogenides based catalytic materials, Mo(W) oxides, borides, phosphides, carbides and nitrides, as well as their hybrids or composites, have also been synthesized with different morphologies, structures, size, porosity and facets by various preparing techniques. The materials presents versatile and excellent catalytic performance in many types of chemical reactions.

Mo(W) carbides and nitrides possess similar properties of noble metals, thus have been investigated in various hydrogenation which is usually catalyzed by noble metals. A lot of studies demonstrated their excellent ability in hydrogenation of CO, CO_2 and aromatics, high alcohol production from syngas etc. However, some disadvantages, such as stability and selectivity, must be solved before they can be applicable in petroleum and chemical industries.

Mo(W) oxide based materials have been employed in a series of oxidation and oxidation dehydrogenation(ODH) reaction to produce bulky and fine chemicals, such as ODH of ethane, ammoxidation of light alkanes and alkenes, selective oxidation of propene to acrylic acid and conversion of glycerol to acrylic acid. These conversions represent ten millions of tons of market demand.

Synergy of Mo(W) oxides with other transition metals (V, Te and Nb) give rise to the M1 active phase which dominates the selectivity and activity of oxidation reaction. The conversion of glycerol to acrylic acid represents one of the most attractive biomass to biochemical processes, and Mo(W) oxide two-functional catalyst is the most promising candidates.

Mo(W) oxide solid acids are recyclable, easy to separate, and free from catalyst waste. They constitute a system of materials that possess thermal stability and strong acidity for many important reactions, such as dehydration, isomerization, alkylation, acylation, esterification, hydration, and hydrolysis. The acidity can be tuned by using mixed oxides, such as ZrO_2-WO_3, ZrO_2-MoO_3, Nb_2O_5-WO_3, and Nb_2O_5-MoO_3. The acid catalysis is strongly influenced by the structures, such as strain, degree of polymerization, coordination patterns of oxide species. And acid strength and acid amount of Mo(W) based oxides greatly affects catalysis reactions.

Mo(W) oxide complex is identified a promising catalyst for metathesis of alkenes to obtain high value of polymer monomers. In metathesis of propene, Mo oxides supported on acidic materials (e.g., silica-alumina or SBA-15) and WO_x supported on SiO_2 are demonstrated to be active.

Considerable worldwide interest exists in discovering renewable energy sources that can substitute for fossil fuels. Mo(W) carbide catalyzes biodiesel production from fats and plant oils, especially nonedible or waste feedstocks, via deoxygenation (DO), and Mo(W) oxide catalyzes ester exchange of glycerides with ethanol and methanol. Now the obtained results are only initial but promising. The main challenge for application is the producing cost and feedstock available. Catalysts based on Mo(W) carbide, nitride, phosphide, boride and oxide have been demonstrated promising performances in a lot of reactions. And they can play important roles in in the sustainable energy and economy, such as CO_2 conversion, water splitting and biomass conversion. But many researches are still required for the application of these catalysts for industry processes.

2. THE PERSPECTIVE

The discovery of potential of graphene have intrigued a surge research on Mo(W) dichalcogenides. To date the significant catalytic potential of 2D Mo(W) based materials have been demonstrated, however, there are considerable challenges in transferring this potential into practical technologies. One of challenge is the scalable production of these 2D materials with efficient and low-cost synthetic techniques. The mechanical exfoliation of bulk crystals or arduous layer-by-layer stacking of multiple layers are clearly not implementable for large practical preparation of catalysts. The thermal CVD process may be only fit for production of fine chemical owing to the low productivity. Large-scale shear-exfoliation of molybdenum disulfide nanosheets has been demonstrated in pioneering work. But improvement in efficiency and productivity must be achieved for industrial application.

The techniques of heterostructure architecture and intercalation tuning of 2D Mo(W) dichalcogenides have constituted a vast pool for synthesizing catalysts with novel functions. The tunable electronic structure through electrochemical or chemical intercalation makes 2D Mo(W) dichalcogenides very attractive candidates for catalysis optimization. These bandgap engineering has the large potential to control the catalytic reaction at atom levels.

The catalytic explorations of 2D layer Mo(W) dichalcogenides are focused on the HER and CO_2 reduction. The rational catalyst design is by searching for the formation of the metastable "high-energy sites" which may provide much higher catalytic activity. Now the active sites are mainly attributed to the Mo or S edge, while the base plane is deemed inert. Activating and optimizing Mo(W) dichalcogenides basal planes for hydrogen evolution may be achieved through the formation of strained sulphur vacancies and defects. Another method is doping metals on edge or base plane, which can give rise to new active sites. Now there are huge opportunities to produce many more fuels or chemicals catalyzed by these 2D type materials, which however needs a lot of investigations and efforts from fundamental researchers and technique engineers.

Concerning the environment has urged the intensive research on the hydrotreatment of $Mo(W)S_2$ based catalysts. But ultra-deep desulfurization usually influences the combustion properties of transport fuels. Hydrogenation selectivity now become more important in development of new catalysts. The coupling and dissociation between desulfurization and hydrogenation will be the main concerning in the hydrotreatment of diesel and gasoline. This needs to discern the active sites for desulfurization and hydrogenation on the $Mo(W)S_2$ based catalysts. Although the "brim" sites of $Co(Ni)MoS_2$ or MoS_2 clusters are proposed to be the hydrogenation centers which are however very close to the active edge for desulfurization reaction. Thus the dissociation of them may be difficult. We recently found by DFT calculation that the DBT and 4,6-DMDBT can adsorb on the edge sites with the aromatic plane vertical to the MoS_2 plane. And the hydrogenation and desulfurization may occur at this adsorption configuration simultaneous. This new discovery suggest a new reaction mechanism which can help to design the newly HDS catalysts.

Although 2D Mo(W) dichalcogenides recently attracts the extensive attention from diversity fields, other Mo(W) based materials should not be neglected because they can catalyze much more reactions, including a lot of oxidation-reduction and acid-base reactions. Carbide and nitride are high active for hydrogenation. The main problems are stability and selectivity. Robust application of Mo(W) oxide based catalysts for selective oxidation need to solve the challenge of selectivity, especially the combustion of feed, intermediate and products.

During the past few decades, "green chemistry" seeks the minimization of adverse effects of chemicals on the environment and human health. One important approach is the development of solid acid catalysts to replace liquid acids, such as sulfuric acid and HF. Liquid acids have several operational problems, such as neutralization, costly and inefficient separation of catalysts from the products, and disposal of the acid catalysts as waste. Mo(W) oxide as solid acid catalyst show interesting potential in industrial application. But the deactivation is the main problem. Coke is easy to form on the catalyst surface, especially on strong acidity sites. Thus improvement of stability and regenerability is the key for commercial application.

Mankind is currently confronting a host of problems related to energy and environment. Among all pollutant gases in the earth atmosphere, CO_2 plays a key role due to its greenhouse gas effect. The unprecedented high CO_2 concentration in atmosphere has been traced mostly to vast emissions from the combustion of carbonaceous fuels, such as coal, oil, and natural gas. One of solving ways is recycling CO_2 gas which can be fulfilled through hydrogenation of CO_2 to methanol and higher alcohols or through the reverse water gas shift reaction to CO. But CO_2 usage is cumbersome due to the challenges associated to the high chemical stability and the difficult activation of this compound. Electric or thermal reduction of CO_2 by noble metal free catalysts is promising. Mo(W) based catalysts represents one of the most potential candidates.

It is urgent to find clean and renewable energy sources to replace fossil energy. Hydrogen produced from splitting water by electrochemical or photochemical process has shown a lot of advantages as a new type of renewable and clean energy, and the important feed in refining and chemical industry. To date, Pt has been recognized as one of the most active catalyst for the hydrogen evolution reaction (HER). However, its high cost and scarce reserves prevents the massive application. Therefore searching the no noble metal HER catalysts can largely remit the environmental and energy issues. Owing to

Summary and Perspectives

the similar electronic structure as Pt-group metals, W and Mo compounds, such as sulfides, phosphides, boride, carbides and nitrides have received increasing attention and intensive investigation. These emerging materials could exhibit outstanding catalytic performance for HER. There are rooms to further enhance their properties by engineering the specific nanostructures. Another promising alternative is the light-driven water splitting where WO_3 have been explored to carry on oxygen-evolving reaction.

Now theoretical description of heterogeneous reactions has developed quickly, with the advances of theoretical methods such as density functional theory. It has now becoming routine to calculate the electronic structures of the catalysts as well as adsorption, surface migration, and surface elementary reactions, to understand the fundamental catalytic properties of the materials. Density functional theory (DFT) have emerged recently as a powerful tool to investigation the structure of the catalysts and the mechanism of the catalytic reactions.

Regarding the molybdenum based catalysts, The theory reveal to the 0D molybdenum carbide (Mo_2C) nanoparticle, the 2D molybdenum sulfide (MoS_2) as well as molybdenum nitride (MoN_2) have help us understand their structures, electronic properties and the mechanism of catalytic reactions. We are now close to a stage when high quality catalysts can be predicted based on theoretical structure-activity relationship, with the electronic structure as a bridge. Moreover, the structures and the mechanism of the real size highly-dispersed catalyst particles in the working conditions can be understood. Tailing molybdenum-based catalysts based on theoretical insights will enable more accurate and efficient development of these catalysts for HDS and many other industrial applications.

The combination of experiment with theoretical calculation can further provide insight into the structure sensitivity, size effects and active sites of reactions. The research can investigate the dynamics in the molecule interaction at edges, corners, and vacancies together with the diffusion of reaction intermediates, which will be important for understanding the HDS and hydrogenation reaction in full detail. These can reveal how structural and morphological changes lead to the changes in the catalytic activity and selectivity. In situ high-resolution transmission electron microscopy may be applied to supported $Mo(W)S_2$ based catalysts. STM can be a useful tool in the investigation of HDS catalysis thanks to that it is applicable under high-temperature and pressure conditions. The so-called high-speed STM may also become an important tool, which provides the analysis of surface dynamics on the atomic-scale. Ambient pressure XPS operating at 1-10 mbar pressure is also becoming available at some large synchrotron facilities, which can study the chemical state of Mo and Co access to catalytic working conditions. The impressive advance in characterization techniques and theory calculation will assist our design and fabrication of the Mo(W)S_2 based catalysts at the atomic level.

In recent decades, catalyst development has transformed from the predominantly empirical test to the situation where it is possible to control the catalytic activity, selectivity and stability via characterizations of the atomic-scale structure and electronic properties. By control of the active centers from atomic to the mesoscale level, more selective and active 2D layer Mo(W)S_2 based catalysts can be made by the combination of the rich fundamental knowledge with the cutting edge synthetic techniques.

Development of catalyst which can sufficiently utilize the active atoms to achieve high activity and selectivity will be our ultimate objectives. These however must be based on the deep understanding of the reaction mechanism, active structures and process engineering. And these are becoming possible with the fast advances in fundamental research and commercial application in petroleum and chemical industries.

Related Readings

To continue IGI Global's long-standing tradition of advancing innovation through emerging research, please find below a compiled list of recommended IGI Global book chapters and journal articles in the areas of oxide-based catalysts, nanosheets, and wet chemical method. These related readings will provide additional information and guidance to further enrich your knowledge and assist you with your own research.

Abu Bakar, W. A., Abdullah, W. N., Ali, R., & Mokhtar, W. N. (2016). Polymolybdate Supported Nano Catalyst for Desulfurization of Diesel. In T. Saleh (Ed.), *Applying Nanotechnology to the Desulfurization Process in Petroleum Engineering* (pp. 263–280). Hershey, PA: IGI Global. doi:10.4018/978-1-4666-9545-0.ch009

Ahmad, W. (2016). Sulfur in Petroleum: Petroleum Desulfurization Techniques. In T. Saleh (Ed.), *Applying Nanotechnology to the Desulfurization Process in Petroleum Engineering* (pp. 1–52). Hershey, PA: IGI Global. doi:10.4018/978-1-4666-9545-0.ch001

Aïssa, B., & Khayyat, M. M. (2014). Self-Healing Materials Systems as a Way for Damage Mitigation in Composites Structures Caused by Orbital Space Debris. In M. Bououdina & J. Davim (Eds.), *Handbook of Research on Nanoscience, Nanotechnology, and Advanced Materials* (pp. 1–25). Hershey, PA: IGI Global. doi:10.4018/978-1-4666-5824-0.ch001

Akbari, E., Buntat, Z., Ahmadi, M. T., Karimi, H., & Khaledian, M. (2017). GAS Sensor Modelling and Simulation. In M. Ahmadi, R. Ismail, & S. Anwar (Eds.), *Handbook of Research on Nanoelectronic Sensor Modeling and Applications* (pp. 70–116). Hershey, PA: IGI Global. doi:10.4018/978-1-5225-0736-9.ch004

Akbari, E., Enzevaee, A., Karimi, H., Ahmadi, M. T., & Buntat, Z. (2017). Graphene-Based Gas Sensor Theoretical Framework. In M. Ahmadi, R. Ismail, & S. Anwar (Eds.), *Handbook of Research on Nanoelectronic Sensor Modeling and Applications* (pp. 117–149). Hershey, PA: IGI Global. doi:10.4018/978-1-5225-0736-9.ch005

Al-Najar, B. T., & Bououdina, M. (2016). Bioinspired Nanoparticles for Efficient Drug Delivery System. In M. Bououdina (Ed.), *Emerging Research on Bioinspired Materials Engineering* (pp. 69–103). Hershey, PA: IGI Global. doi:10.4018/978-1-4666-9811-6.ch003

AlMegren, H. A., Gonzalez-Cortes, S., Huang, Y., Chen, H., Qian, Y., Alkinany, M., & Xiao, T. et al. (2016). Preparation of Deep Hydrodesulfurzation Catalysts for Diesel Fuel using Organic Matrix Decomposition Method. In H. Al-Megren & T. Xiao (Eds.), *Petrochemical Catalyst Materials, Processes, and Emerging Technologies* (pp. 216–253). Hershey, PA: IGI Global. doi:10.4018/978-1-4666-9975-5.ch009

Alshammari, A., Kalevaru, V. N., Bagabas, A., & Martin, A. (2016). Production of Ethylene and its Commercial Importance in the Global Market. In H. Al-Megren & T. Xiao (Eds.), *Petrochemical Catalyst Materials, Processes, and Emerging Technologies* (pp. 82–115). Hershey, PA: IGI Global. doi:10.4018/978-1-4666-9975-5.ch004

Anwar, S. (2017). Wireless Nanosensor Networks: Prospects and Challenges. In M. Ahmadi, R. Ismail, & S. Anwar (Eds.), *Handbook of Research on Nanoelectronic Sensor Modeling and Applications* (pp. 505–511). Hershey, PA: IGI Global. doi:10.4018/978-1-5225-0736-9.ch017

Arafat, M. T., & Li, X. (2016). Functional Coatings for Bone Tissue Engineering. In A. Zuzuarregui & M. Morant-Miñana (Eds.), *Research Perspectives on Functional Micro- and Nanoscale Coatings* (pp. 240–264). Hershey, PA: IGI Global. doi:10.4018/978-1-5225-0066-7.ch009

Arokiyaraj, S., Saravanan, M., Bharanidharan, R., Islam, V. I., Bououdina, M., & Vincent, S. (2016). Green Synthesis of Metallic Nanoparticles Using Plant Compounds and Their Applications: Metallic Nanoparticles Synthesis Using Plants. In M. Bououdina (Ed.), *Emerging Research on Bioinspired Materials Engineering* (pp. 1–34). Hershey, PA: IGI Global. doi:10.4018/978-1-4666-9811-6.ch001

Related Readings

Balachandran, P. V., & Rondinelli, J. M. (2016). Informatics-Based Approaches for Accelerated Discovery of Functional Materials. In S. Datta & J. Davim (Eds.), *Computational Approaches to Materials Design: Theoretical and Practical Aspects* (pp. 192–223). Hershey, PA: IGI Global. doi:10.4018/978-1-5225-0290-6.ch007

Bamufleh, H. S., Noureldin, M. M., & El-Halwagi, M. M. (2016). Sustainable Process Integration in the Petrochemical Industries. In H. Al-Megren & T. Xiao (Eds.), *Petrochemical Catalyst Materials, Processes, and Emerging Technologies* (pp. 150–163). Hershey, PA: IGI Global. doi:10.4018/978-1-4666-9975-5.ch006

Banerjee, S., Gautam, R. K., Gautam, P. K., Jaiswal, A., & Chattopadhyaya, M. C. (2016). Recent Trends and Advancement in Nanotechnology for Water and Wastewater Treatment: Nanotechnological Approach for Water Purification. In A. Khitab & W. Anwar (Eds.), *Advanced Research on Nanotechnology for Civil Engineering Applications* (pp. 208–252). Hershey, PA: IGI Global. doi:10.4018/978-1-5225-0344-6.ch007

Barbero, C. A., & Yslas, E. I. (2017). Ecotoxicity Effects of Nanomaterials on Aquatic Organisms: Nanotoxicology of Materials on Aquatic Organisms. In S. Joo (Ed.), *Applying Nanotechnology for Environmental Sustainability* (pp. 330–351). Hershey, PA: IGI Global. doi:10.4018/978-1-5225-0585-3.ch014

Barbhuiya, S. (2014). Applications of Nanomaterials in Construction Industry. In M. Bououdina & J. Davim (Eds.), *Handbook of Research on Nanoscience, Nanotechnology, and Advanced Materials* (pp. 164–175). Hershey, PA: IGI Global. doi:10.4018/978-1-4666-5824-0.ch008

Barbu, M. C., Reh, R., & Çavdar, A. D. (2014). Non-Wood Lignocellulosic Composites. In A. Aguilera & J. Davim (Eds.), *Research Developments in Wood Engineering and Technology* (pp. 281–319). Hershey, PA: IGI Global. doi:10.4018/978-1-4666-4554-7.ch008

Barbu, M. C., Reh, R., & Irle, M. (2014). Wood-Based Composites. In A. Aguilera & J. Davim (Eds.), *Research Developments in Wood Engineering and Technology* (pp. 1–45). Hershey, PA: IGI Global. doi:10.4018/978-1-4666-4554-7.ch001

Bashir, R., & Chisti, H. (2015). Nanotechnology for Environmental Control and Remediation. In M. Shah, M. Bhat, & J. Davim (Eds.), *Nanotechnology Applications for Improvements in Energy Efficiency and Environmental Management* (pp. 156–183). Hershey, PA: IGI Global. doi:10.4018/978-1-4666-6304-6.ch006

Bayir, E., Bilgi, E., & Urkmez, A. S. (2014). Implementation of Nanoparticles in Cancer Therapy. In M. Bououdina & J. Davim (Eds.), *Handbook of Research on Nanoscience, Nanotechnology, and Advanced Materials* (pp. 447–491). Hershey, PA: IGI Global. doi:10.4018/978-1-4666-5824-0.ch018

Benjamin, S., Unni, K. N., Priji, P., & Wright, A. G. (2017). Biogenesis of Conjugated Linoleic Acids. In S. Benjamin (Ed.), *Examining the Development, Regulation, and Consumption of Functional Foods* (pp. 1–28). Hershey, PA: IGI Global. doi:10.4018/978-1-5225-0607-2.ch001

Berger, L. (2015). Tribology of Thermally Sprayed Coatings in the Al_2O_3-Cr_2O_3-TiO_2 System. In M. Roy & J. Davim (Eds.), *Thermal Sprayed Coatings and their Tribological Performances* (pp. 227–267). Hershey, PA: IGI Global. doi:10.4018/978-1-4666-7489-9.ch008

Bhat, M. A., Nayak, B. K., Nanda, A., & Lone, I. H. (2015). Nanotechnology, Metal Nanoparticles, and Biomedical Applications of Nanotechnology. In M. Shah, M. Bhat, & J. Davim (Eds.), *Nanotechnology Applications for Improvements in Energy Efficiency and Environmental Management* (pp. 116–155). Hershey, PA: IGI Global. doi:10.4018/978-1-4666-6304-6.ch005

Bhutto, A. W., Abro, R., Abbas, T., Yu, G., & Chen, X. (2016). Desulphurization of Fuel Oils Using Ionic Liquids. In H. Al-Megren & T. Xiao (Eds.), *Petrochemical Catalyst Materials, Processes, and Emerging Technologies* (pp. 254–284). Hershey, PA: IGI Global. doi:10.4018/978-1-4666-9975-5.ch010

Bodratti, A. M., He, Z., Tsianou, M., Cheng, C., & Alexandridis, P. (2015). Product Design Applied to Formulated Products: A Course on Their Design and Development that Integrates Knowledge of Materials Chemistry, (Nano) Structure and Functional Properties. *International Journal of Quality Assurance in Engineering and Technology Education*, 4(3), 21–43. doi:10.4018/IJQAETE.2015070102

Bogataj, D., & Drobne, D. (2017). Control of Perishable Goods in Cold Logistic Chains by Bionanosensors. In S. Joo (Ed.), *Applying Nanotechnology for Environmental Sustainability* (pp. 376–402). Hershey, PA: IGI Global. doi:10.4018/978-1-5225-0585-3.ch016

Related Readings

Bolboacă, S. D., & Jäntschi, L. (2017). Characteristic Polynomial in Assessment of Carbon-Nano Structures. In M. Putz & M. Mirica (Eds.), *Sustainable Nanosystems Development, Properties, and Applications* (pp. 122–147). Hershey, PA: IGI Global. doi:10.4018/978-1-5225-0492-4.ch004

Bouloudenine, M., & Bououdina, M. (2016). Toxic Effects of Engineered Nanoparticles on Living Cells. In M. Bououdina (Ed.), *Emerging Research on Bioinspired Materials Engineering* (pp. 35–68). Hershey, PA: IGI Global. doi:10.4018/978-1-4666-9811-6.ch002

Boumaza, N., Benouaz, T., & Goumri-Said, S. (2014). Understanding the Numerical Resolution of Perturbed Soliton Propagation in Single Mode Optical Fiber. In M. Bououdina & J. Davim (Eds.), *Handbook of Research on Nanoscience, Nanotechnology, and Advanced Materials* (pp. 492–504). Hershey, PA: IGI Global. doi:10.4018/978-1-4666-5824-0.ch019

Brimmo, A., & Emziane, M. (2014). Carbon Nanotubes for Photovoltaics. In M. Bououdina & J. Davim (Eds.), *Handbook of Research on Nanoscience, Nanotechnology, and Advanced Materials* (pp. 268–311). Hershey, PA: IGI Global. doi:10.4018/978-1-4666-5824-0.ch012

Brunetti, A., Sellaro, M., Drioli, E., & Barbieri, G. (2016). Membrane Engineering and its Role in Oil Refining and Petrochemical Industry. In H. Al-Megren & T. Xiao (Eds.), *Petrochemical Catalyst Materials, Processes, and Emerging Technologies* (pp. 116–149). Hershey, PA: IGI Global. doi:10.4018/978-1-4666-9975-5.ch005

Choi, H., Son, M., Bae, J., & Choi, H. (2017). Nanotechnology in Engineered Membranes: Innovative Membrane Material for Water-Energy Nexus. In S. Joo (Ed.), *Applying Nanotechnology for Environmental Sustainability* (pp. 50–71). Hershey, PA: IGI Global. doi:10.4018/978-1-5225-0585-3.ch003

del Valle-Zermeño, R., Chimenos, J. M., & Formosa, J. (2016). Flue Gas Desulfurization: Processes and Technologies. In T. Saleh (Ed.), *Applying Nanotechnology to the Desulfurization Process in Petroleum Engineering* (pp. 337–377). Hershey, PA: IGI Global. doi:10.4018/978-1-4666-9545-0.ch011

Deng, Y., & Liu, S. (2016). Catalysis with Room Temperature Ionic Liquids Mediated Metal Nanoparticles. In H. Al-Megren & T. Xiao (Eds.), *Petrochemical Catalyst Materials, Processes, and Emerging Technologies* (pp. 285–329). Hershey, PA: IGI Global. doi:10.4018/978-1-4666-9975-5.ch011

Dimitratos, N., Villa, A., Chan-Thaw, C. E., Hammond, C., & Prati, L. (2016). Valorisation of Glycerol to Fine Chemicals and Fuels. In H. Al-Megren & T. Xiao (Eds.), *Petrochemical Catalyst Materials, Processes, and Emerging Technologies* (pp. 352–384). Hershey, PA: IGI Global. doi:10.4018/978-1-4666-9975-5.ch013

Dixit, A. K., & Awasthi, R. (2015). EDM Process Parameters Optimization for Al-TiO2 Nano Composite. *International Journal of Materials Forming and Machining Processes, 2*(2), 17–30. doi:10.4018/IJMFMP.2015070102

Elrawemi, M., Blunt, L., Fleming, L., Sweeney, F., Robbins, D., & Bird, D. (2015). Metrology of Al2O3 Barrier Film for Flexible CIGS Solar Cells. *International Journal of Energy Optimization and Engineering, 4*(4), 46–60. doi:10.4018/IJEOE.2015100104

Emam, A. N., Mansour, A. S., Girgis, E., & Mohamed, M. B. (2017). Hybrid Nanostructures: Synthesis and Physicochemical Characterizations of Plasmonic Nanocomposites. In S. Joo (Ed.), *Applying Nanotechnology for Environmental Sustainability* (pp. 231–275). Hershey, PA: IGI Global. doi:10.4018/978-1-5225-0585-3.ch011

Emam, A. N., Mansour, A. S., Girgis, E., & Mohamed, M. B. (2017). Hybrid Plasmonic Nanostructures: Environmental Impact and Applications. In S. Joo (Ed.), *Applying Nanotechnology for Environmental Sustainability* (pp. 276–293). Hershey, PA: IGI Global. doi:10.4018/978-1-5225-0585-3.ch012

Emziane, M., & Yoosuf, R. (2014). In2X3 (X=S, Se, Te) Semiconductor Thin Films: Fabrication, Properties, and Applications. In M. Bououdina & J. Davim (Eds.), *Handbook of Research on Nanoscience, Nanotechnology, and Advanced Materials* (pp. 226–267). Hershey, PA: IGI Global. doi:10.4018/978-1-4666-5824-0.ch011

Fathi, A., Azizian, S., & Sharifan, N. (2017). Sensors and Amplifiers: Sensor Output Signal Amplification Systems. In M. Ahmadi, R. Ismail, & S. Anwar (Eds.), *Handbook of Research on Nanoelectronic Sensor Modeling and Applications* (pp. 423–504). Hershey, PA: IGI Global. doi:10.4018/978-1-5225-0736-9.ch016

Fathi, A., & Hassanzadazar, M. (2017). CNT as a Sensor Platform. In M. Ahmadi, R. Ismail, & S. Anwar (Eds.), *Handbook of Research on Nanoelectronic Sensor Modeling and Applications* (pp. 1–18). Hershey, PA: IGI Global. doi:10.4018/978-1-5225-0736-9.ch001

Related Readings

Ferbinteanu, M., Ramanantoanina, H., & Cimpoesu, F. (2017). Case Studies in the Challenge of Properties Design at Nanoscale: Bonding Mechanisms and causal Relationship. In M. Putz & M. Mirica (Eds.), *Sustainable Nanosystems Development, Properties, and Applications* (pp. 148–184). Hershey, PA: IGI Global. doi:10.4018/978-1-5225-0492-4.ch005

Fernández, S. M., & Marijuan, A. G. (2016). High-Tech Applications of Functional Coatings: Functional Coatings and Photovoltaic. In A. Zuzuarregui & M. Morant-Miñana (Eds.), *Research Perspectives on Functional Micro- and Nanoscale Coatings* (pp. 289–317). Hershey, PA: IGI Global. doi:10.4018/978-1-5225-0066-7.ch011

Fjodorova, N., Novic, M., Diankova, T., & Ostanen, A. (2016). The Nano-Sized TiO2 Dispersions for Mass Coloration of Polyimide Fibers: The Nano-Sized TiO2 for Mass Coloration. *Journal of Nanotoxicology and Nanomedicine*, *1*(1), 29–44. doi:10.4018/JNN.2016010103

Florea, L., Diamond, D., & Benito-Lopez, F. (2016). Opto-Smart Systems in Microfluidics. In A. Zuzuarregui & M. Morant-Miñana (Eds.), *Research Perspectives on Functional Micro- and Nanoscale Coatings* (pp. 265–288). Hershey, PA: IGI Global. doi:10.4018/978-1-5225-0066-7.ch010

Gaines, T. W., Williams, K. R., & Wagener, K. B. (2016). ADMET: Functionalized Polyolefins. In H. Al-Megren & T. Xiao (Eds.), *Petrochemical Catalyst Materials, Processes, and Emerging Technologies* (pp. 1–21). Hershey, PA: IGI Global. doi:10.4018/978-1-4666-9975-5.ch001

Galian, R. E., & Pérez-Prieto, J. (2016). Synergism at the Nanoscale: Photoactive Semiconductor Nanoparticles and their Organic Ligands. In A. Zuzuarregui & M. Morant-Miñana (Eds.), *Research Perspectives on Functional Micro- and Nanoscale Coatings* (pp. 42–77). Hershey, PA: IGI Global. doi:10.4018/978-1-5225-0066-7.ch003

Gaspar-Cunha, A., & Covas, J. A. (2014). An Engineering Scale-Up Approach using Multi-Objective Optimization. *International Journal of Natural Computing Research*, *4*(1), 17–30. doi:10.4018/ijncr.2014010102

Gaurina-Medjimurec, N., & Pasic, B. (2014). Risk Due to Pipe Sticking. In D. Matanovic, N. Gaurina-Medjimurec, & K. Simon (Eds.), *Risk Analysis for Prevention of Hazardous Situations in Petroleum and Natural Gas Engineering* (pp. 47–72). Hershey, PA: IGI Global. doi:10.4018/978-1-4666-4777-0.ch003

Gaurina-Medjimurec, N., & Pasic, B. (2014). Risk Due to Wellbore Instability. In D. Matanovic, N. Gaurina-Medjimurec, & K. Simon (Eds.), *Risk Analysis for Prevention of Hazardous Situations in Petroleum and Natural Gas Engineering* (pp. 23–46). Hershey, PA: IGI Global. doi:10.4018/978-1-4666-4777-0.ch002

Ge, H., Tang, M., & Wen, X. (2016). Ni/ZnO Nano Sorbent for Reactive Adsorption Desulfurization of Refinery Oil Streams. In T. Saleh (Ed.), *Applying Nanotechnology to the Desulfurization Process in Petroleum Engineering* (pp. 216–239). Hershey, PA: IGI Global. doi:10.4018/978-1-4666-9545-0.ch007

Golden, T. D., Tientong, J., & Mohamed, A. M. (2016). Electrodeposition of Nickel-Molybdenum (Ni-Mo) Alloys for Corrosion Protection in Harsh Environments. In A. Zuzuarregui & M. Morant-Miñana (Eds.), *Research Perspectives on Functional Micro- and Nanoscale Coatings* (pp. 369–395). Hershey, PA: IGI Global. doi:10.4018/978-1-5225-0066-7.ch014

Gopal, S., & Al-Hazmi, M. H. (2016). Advances in Catalytic Technologies for Selective Oxidation of Lower Alkanes. In H. Al-Megren & T. Xiao (Eds.), *Petrochemical Catalyst Materials, Processes, and Emerging Technologies* (pp. 22–52). Hershey, PA: IGI Global. doi:10.4018/978-1-4666-9975-5.ch002

Gupta, A. K., Dey, A., & Mukhopadhyay, A. K. (2016). Micromechanical and Finite Element Modeling for Composites. In S. Datta & J. Davim (Eds.), *Computational Approaches to Materials Design: Theoretical and Practical Aspects* (pp. 101–162). Hershey, PA: IGI Global. doi:10.4018/978-1-5225-0290-6.ch005

Han, B., Liu, W., & Zhao, D. (2017). In-Situ Oxidative Degradation of Emerging Contaminants in Soil and Groundwater Using a New Class of Stabilized MnO2 Nanoparticles. In S. Joo (Ed.), *Applying Nanotechnology for Environmental Sustainability* (pp. 112–136). Hershey, PA: IGI Global. doi:10.4018/978-1-5225-0585-3.ch006

Holberg, S. (2016). Non-Hydrolyzed Resins for Organic-Inorganic Hybrid Coatings: Functional Coating Films by Moisture Curing. In A. Zuzuarregui & M. Morant-Miñana (Eds.), *Research Perspectives on Functional Micro- and Nanoscale Coatings* (pp. 105–135). Hershey, PA: IGI Global. doi:10.4018/978-1-5225-0066-7.ch005

Related Readings

Hrncevic, L. (2014). Petroleum Industry Environmental Performance and Risk. In D. Matanovic, N. Gaurina-Medjimurec, & K. Simon (Eds.), *Risk Analysis for Prevention of Hazardous Situations in Petroleum and Natural Gas Engineering* (pp. 358–387). Hershey, PA: IGI Global. doi:10.4018/978-1-4666-4777-0.ch016

Hu, Y., Peng, Y., Liu, W., Zhao, D., & Fu, J. (2017). Removal of Emerging Contaminants from Water and Wastewater Using Nanofiltration Technology. In S. Joo (Ed.), *Applying Nanotechnology for Environmental Sustainability* (pp. 72–91). Hershey, PA: IGI Global. doi:10.4018/978-1-5225-0585-3.ch004

Huirache-Acuña, R., Alonso-Nuñez, G., Rivera-Muñoz, E. M., Gutierrez, O., & Pawelec, B. (2016). Trimetallic Sulfide Catalysts for Hydrodesulfurization. In T. Saleh (Ed.), *Applying Nanotechnology to the Desulfurization Process in Petroleum Engineering* (pp. 240–262). Hershey, PA: IGI Global. doi:10.4018/978-1-4666-9545-0.ch008

Huyen, P. T. (2017). Clay Minerals Converted to Porous Materials and Their Application: Challenge and Perspective. In T. Kobayashi (Ed.), *Applied Environmental Materials Science for Sustainability* (pp. 141–164). Hershey, PA: IGI Global. doi:10.4018/978-1-5225-1971-3.ch007

Jayavarthanan, R., Nanda, A., & Bhat, M. A. (2017). The Impact of Nanotechnology on Environment. In B. Nayak, A. Nanda, & M. Bhat (Eds.), *Integrating Biologically-Inspired Nanotechnology into Medical Practice* (pp. 153–193). Hershey, PA: IGI Global. doi:10.4018/978-1-5225-0610-2.ch007

Julião, D., Ribeiro, S., de Castro, B., Cunha-Silva, L., & Balula, S. S. (2016). Polyoxometalates-Based Nanocatalysts for Production of Sulfur-Free Diesel. In T. Saleh (Ed.), *Applying Nanotechnology to the Desulfurization Process in Petroleum Engineering* (pp. 426–458). Hershey, PA: IGI Global. doi:10.4018/978-1-4666-9545-0.ch014

Kamaja, C. K., Rajaperumal, M., Boukherroub, R., & Shelke, M. V. (2014). Silicon Nanostructures-Graphene Nanocomposites: Efficient Materials for Energy Conversion and Storage. In M. Bououdina & J. Davim (Eds.), *Handbook of Research on Nanoscience, Nanotechnology, and Advanced Materials* (pp. 176–195). Hershey, PA: IGI Global. doi:10.4018/978-1-4666-5824-0.ch009

Related Readings

Kannoorpatti, K., & Surovtseva, D. (2015). Integrating Industry Research in Pedagogical Practice: A Case of Teaching Microbial Corrosion in Wet Tropics. In H. Lim (Ed.), *Handbook of Research on Recent Developments in Materials Science and Corrosion Engineering Education* (pp. 254–272). Hershey, PA: IGI Global. doi:10.4018/978-1-4666-8183-5.ch013

Kanoun, M. B., & Goumri-Said, S. (2014). Theoretical Assessment of the Mechanical, Electronic, and Vibrational Properties of the Paramagnetic Insulating Cerium Dioxide and Investigation of Intrinsic Defects. In M. Bououdina & J. Davim (Eds.), *Handbook of Research on Nanoscience, Nanotechnology, and Advanced Materials* (pp. 431–446). Hershey, PA: IGI Global. doi:10.4018/978-1-4666-5824-0.ch017

Karimi, H., Rahmani, R., Akbari, E., Darabi, A. C., Rahmani, M., Ahmadi, M. T., & Anbari, S. (2017). Modeling of Sensing Layer of Surface Acoustic-Wave-Based Gas Sensors. In M. Ahmadi, R. Ismail, & S. Anwar (Eds.), *Handbook of Research on Nanoelectronic Sensor Modeling and Applications* (pp. 224–243). Hershey, PA: IGI Global. doi:10.4018/978-1-5225-0736-9.ch009

Karimi, H., Rahmani, R., Akbari, E., Rahmani, M., & Ahamdi, M. T. (2017). Optimization of Current-Voltage Characteristics of Graphene-Based Biosensors. In M. Ahmadi, R. Ismail, & S. Anwar (Eds.), *Handbook of Research on Nanoelectronic Sensor Modeling and Applications* (pp. 244–264). Hershey, PA: IGI Global. doi:10.4018/978-1-5225-0736-9.ch010

Kasani, H., Ahmadi, M. T., Khoda-Bakhsh, R., & Ochbelagh, D. R. (2017). Fast Neuron Detection. In M. Ahmadi, R. Ismail, & S. Anwar (Eds.), *Handbook of Research on Nanoelectronic Sensor Modeling and Applications* (pp. 395–422). Hershey, PA: IGI Global. doi:10.4018/978-1-5225-0736-9.ch015

Khalaj, Z., Monajjemi, M., & Diudea, M. V. (2017). Main Allotropes of Carbon: A Brief Review. In M. Putz & M. Mirica (Eds.), *Sustainable Nanosystems Development, Properties, and Applications* (pp. 185–213). Hershey, PA: IGI Global. doi:10.4018/978-1-5225-0492-4.ch006

Khanday, M. F. (2015). Convergence of Nanotechnology and Microbiology. In M. Shah, M. Bhat, & J. Davim (Eds.), *Nanotechnology Applications for Improvements in Energy Efficiency and Environmental Management* (pp. 313–342). Hershey, PA: IGI Global. doi:10.4018/978-1-4666-6304-6.ch011

Related Readings

Khayyat, M. M., & Aïssa, B. (2014). Si-NWs: Major Advances in Synthesis and Applications. In M. Bououdina & J. Davim (Eds.), *Handbook of Research on Nanoscience, Nanotechnology, and Advanced Materials* (pp. 108–130). Hershey, PA: IGI Global. doi:10.4018/978-1-4666-5824-0.ch005

Kiani, M. J., Abadi, M. H., Rahmani, M., Ahmadi, M. T., Harun, F. K., & Bagherifard, K. (2017). Graphene Based-Biosensor: Graphene Based Electrolyte Gated Graphene Field Effect Transistor. In M. Ahmadi, R. Ismail, & S. Anwar (Eds.), *Handbook of Research on Nanoelectronic Sensor Modeling and Applications* (pp. 265–293). Hershey, PA: IGI Global. doi:10.4018/978-1-5225-0736-9.ch011

Kiani, M. J., Abadi, M. H., Rahmani, M., Ahmadi, M. T., Harun, F. K., Hedayat, S., & Yaghoobian, S. (2017). Carbon Materials Based Ion Sensitive Field Effect Transistor (ISFET): The Emerging Potentials of Nanostructured Carbon-Based ISFET with High Sensitivity. In M. Ahmadi, R. Ismail, & S. Anwar (Eds.), *Handbook of Research on Nanoelectronic Sensor Modeling and Applications* (pp. 334–360). Hershey, PA: IGI Global. doi:10.4018/978-1-5225-0736-9.ch013

Krasnoholovets, V. (2017). A Theoretical Study of the Refractive Index of KDP Crystal Doped with TiO2 Nanoparticles. In M. Putz & M. Mirica (Eds.), *Sustainable Nanosystems Development, Properties, and Applications* (pp. 524–534). Hershey, PA: IGI Global. doi:10.4018/978-1-5225-0492-4.ch013

Krishnapriya, K., & Ramesh, M. (2017). Copper and Copper Nanoparticles Induced Hematological Changes in a Freshwater Fish Labeo rohita – A Comparative Study: Copper and Copper Nanoparticle Toxicity to Fish. In S. Joo (Ed.), *Applying Nanotechnology for Environmental Sustainability* (pp. 352–375). Hershey, PA: IGI Global. doi:10.4018/978-1-5225-0585-3.ch015

Kvasnička, V., & Pospíchal, J. (2014). A Study of Replicators and Hypercycles by Hofstadters Typogenetics. *International Journal of Signs and Semiotic Systems*, *3*(1), 10–26. doi:10.4018/ijsss.2014010102

Kwon, K., Li, L., & Kim, D. (2015). Energy Harvesting from Wastewater Using Nanofluidic Reverse Electrodialysis. In L. Mescia, O. Losito, & F. Prudenzano (Eds.), *Innovative Materials and Systems for Energy Harvesting Applications* (pp. 380–411). Hershey, PA: IGI Global. doi:10.4018/978-1-4666-8254-2.ch013

László, I., Zsoldos, I., & Fülep, D. (2017). Self Organizing Carbon Structures: Tight Binding Molecular Dynamics Calculations. In M. Putz & M. Mirica (Eds.), *Sustainable Nanosystems Development, Properties, and Applications* (pp. 46–58). Hershey, PA: IGI Global. doi:10.4018/978-1-5225-0492-4.ch002

Lavanya, K., Durai, M. S., & Iyengar, N. (2016). A Hybrid Model for Rice Disease Diagnosis Using Entropy Based Neuro Genetic Algorithm. *International Journal of Agricultural and Environmental Information Systems*, 7(2), 52–69. doi:10.4018/IJAEIS.2016040103

Li, H., & Harruna, I. (2014). Functionalization of Carbon Nanocomposites with Ruthenium Bipyridine and Terpyridine Complex. In M. Bououdina & J. Davim (Eds.), *Handbook of Research on Nanoscience, Nanotechnology, and Advanced Materials* (pp. 26–61). Hershey, PA: IGI Global. doi:10.4018/978-1-4666-5824-0.ch002

Li, K., & Kobayashi, T. (2017). Ionic Liquids and Poly (Ionic Liquid)s Used as Green Solvent and Ultrasound Responded Materials. In T. Kobayashi (Ed.), *Applied Environmental Materials Science for Sustainability* (pp. 327–346). Hershey, PA: IGI Global. doi:10.4018/978-1-5225-1971-3.ch015

Li, M. (2014). Pharmacokinetics of Polymeric Nanoparticles at Whole Body, Organ, Cell, and Molecule Levels. In M. Bououdina & J. Davim (Eds.), *Handbook of Research on Nanoscience, Nanotechnology, and Advanced Materials* (pp. 146–163). Hershey, PA: IGI Global. doi:10.4018/978-1-4666-5824-0.ch007

López, C. Y., Bueno, J. J., Torres, I. Z., Mendoza-López, M. L., Álvarez, J. E., & Macías, A. H. (2015). Electrophoretical Deposition of Nanotube TiO2 Conglomerates Detached During Ti Anodizing Used for Decomposing Methyl Orange in Water. In S. Soni, A. Salhotra, & M. Suar (Eds.), *Handbook of Research on Diverse Applications of Nanotechnology in Biomedicine, Chemistry, and Engineering* (pp. 477–495). Hershey, PA: IGI Global. doi:10.4018/978-1-4666-6363-3.ch022

Lucas, C. E., & Pueyo, C. L. (2016). Single Source Precursors for Semiconducting Metal Oxide-Based Films. In A. Zuzuarregui & M. Morant-Miñana (Eds.), *Research Perspectives on Functional Micro- and Nanoscale Coatings* (pp. 26–41). Hershey, PA: IGI Global. doi:10.4018/978-1-5225-0066-7.ch002

Related Readings

Marcu, I., Urdă, A., Popescu, I., & Hulea, V. (2017). Layered Double Hydroxides-Based Materials as Oxidation Catalysts. In M. Putz & M. Mirica (Eds.), *Sustainable Nanosystems Development, Properties, and Applications* (pp. 59–121). Hershey, PA: IGI Global. doi:10.4018/978-1-5225-0492-4.ch003

Martin, A., Kalevaru, V. N., & Radnik, J. (2016). Palladium in Heterogeneous Oxidation Catalysis. In H. Al-Megren & T. Xiao (Eds.), *Petrochemical Catalyst Materials, Processes, and Emerging Technologies* (pp. 53–81). Hershey, PA: IGI Global. doi:10.4018/978-1-4666-9975-5.ch003

Matanovic, D. (2014). General Approach to Risk Analysis. In D. Matanovic, N. Gaurina-Medjimurec, & K. Simon (Eds.), *Risk Analysis for Prevention of Hazardous Situations in Petroleum and Natural Gas Engineering* (pp. 1–22). Hershey, PA: IGI Global. doi:10.4018/978-1-4666-4777-0.ch001

McDonald, K. D., Ojo, E. O., & Liebman, J. F. (2017). What Are the Structures of the Octet Rule Obeying All-Carbon Species Cx ($2 \leq x \leq 7$ and Larger x)?: A Pedagogical, Mathematical, and Pictorial Study. In M. Putz & M. Mirica (Eds.), *Sustainable Nanosystems Development, Properties, and Applications* (pp. 1–45). Hershey, PA: IGI Global. doi:10.4018/978-1-5225-0492-4.ch001

Melnyczuk, J. M., & Palchoudhury, S. (2014). Synthesis and Characterization of Iron Oxide Nanoparticles. In M. Bououdina & J. Davim (Eds.), *Handbook of Research on Nanoscience, Nanotechnology, and Advanced Materials* (pp. 89–107). Hershey, PA: IGI Global. doi:10.4018/978-1-4666-5824-0.ch004

Meshginqalam, B., Ahmadi, M. T., Tousi, H. T., Sabatyan, A., & Centeno, A. (2017). Surface Plasmon Resonance-Based Sensor Modeling. In M. Ahmadi, R. Ismail, & S. Anwar (Eds.), *Handbook of Research on Nanoelectronic Sensor Modeling and Applications* (pp. 361–394). Hershey, PA: IGI Global. doi:10.4018/978-1-5225-0736-9.ch014

Mir, S. A., Shah, M. A., Mir, M. M., & Iqbal, U. (2017). New Horizons of Nanotechnology in Agriculture and Food Processing Industry. In B. Nayak, A. Nanda, & M. Bhat (Eds.), *Integrating Biologically-Inspired Nanotechnology into Medical Practice* (pp. 230–258). Hershey, PA: IGI Global. doi:10.4018/978-1-5225-0610-2.ch009

Mischler, S., & Roy, M. (2015). Tribocorrosion of Thermal Sprayed Coatings. In M. Roy & J. Davim (Eds.), *Thermal Sprayed Coatings and their Tribological Performances* (pp. 25–60). Hershey, PA: IGI Global. doi:10.4018/978-1-4666-7489-9.ch002

Mishra, G., Pandey, S., Dutta, A., & Giri, K. (2017). Nanotechnology Applications for Sustainable Crop Production. In S. Joo (Ed.), *Applying Nanotechnology for Environmental Sustainability* (pp. 164–184). Hershey, PA: IGI Global. doi:10.4018/978-1-5225-0585-3.ch008

Misra, R., & Rao, N. N. (2015). Electrochemical Technologies for Industrial Effluent Treatment. In N. Gaurina-Medjimurec (Ed.), *Handbook of Research on Advancements in Environmental Engineering* (pp. 118–146). Hershey, PA: IGI Global. doi:10.4018/978-1-4666-7336-6.ch005

Mitra-Kirtley, S., Mullins, O. C., & Pomerantz, A. E. (2016). Sulfur and Nitrogen Chemical Speciation in Crude Oils and Related Carbonaceous Materials. In T. Saleh (Ed.), *Applying Nanotechnology to the Desulfurization Process in Petroleum Engineering* (pp. 53–83). Hershey, PA: IGI Global. doi:10.4018/978-1-4666-9545-0.ch002

Mohamed, A. M., Abdullah, A. M., Al-Maadeed, M., & Bahgat, A. (2016). Fundamental, Fabrication and Applications of Superhydrophobic Surfaces. In A. Zuzuarregui & M. Morant-Miñana (Eds.), *Research Perspectives on Functional Micro- and Nanoscale Coatings* (pp. 341–368). Hershey, PA: IGI Global. doi:10.4018/978-1-5225-0066-7.ch013

Montalvan-Sorrosa, D., de los Cobos-Vasconcelos, D., & Gonzalez-Sanchez, A. (2016). Nanotechnology Applied to the Biodesulfurization of Fossil Fuels and Spent Caustic Streams. In T. Saleh (Ed.), *Applying Nanotechnology to the Desulfurization Process in Petroleum Engineering* (pp. 378–389). Hershey, PA: IGI Global. doi:10.4018/978-1-4666-9545-0.ch012

Moor, K., Snow, S., & Kim, J. (2017). Light Sensitized Disinfection with Fullerene. In S. Joo (Ed.), *Applying Nanotechnology for Environmental Sustainability* (pp. 137–163). Hershey, PA: IGI Global. doi:10.4018/978-1-5225-0585-3.ch007

Nogueira, V. I., Gavina, A., Bouguerra, S., Andreani, T., Lopes, I., Rocha-Santos, T., & Pereira, R. (2017). Ecotoxicity and Toxicity of Nanomaterials with Potential for Wastewater Treatment Applications. In S. Joo (Ed.), *Applying Nanotechnology for Environmental Sustainability* (pp. 294–329). Hershey, PA: IGI Global. doi:10.4018/978-1-5225-0585-3.ch013

Related Readings

Nunnelley, K. G., & Smith, J. A. (2017). Nanotechnology for Filtration-Based Point-of-Use Water Treatment: A Review of Current Understanding. In S. Joo (Ed.), *Applying Nanotechnology for Environmental Sustainability* (pp. 27–49). Hershey, PA: IGI Global. doi:10.4018/978-1-5225-0585-3.ch002

Obayya, S., Areed, N. F., Hameed, M. F., & Abdelrazik, M. H. (2015). Optical Nano-Antennas for Energy Harvesting. In L. Mescia, O. Losito, & F. Prudenzano (Eds.), *Innovative Materials and Systems for Energy Harvesting Applications* (pp. 26–62). Hershey, PA: IGI Global. doi:10.4018/978-1-4666-8254-2.ch002

Ogunlaja, A. S., & Tshentu, Z. R. (2016). Molecularly Imprinted Polymer Nanofibers for Adsorptive Desulfurization. In T. Saleh (Ed.), *Applying Nanotechnology to the Desulfurization Process in Petroleum Engineering* (pp. 281–336). Hershey, PA: IGI Global. doi:10.4018/978-1-4666-9545-0.ch010

Pakseresht, A., Rahimipour, M., Alizadeh, M., Hadavi, S., & Shahbazkhan, A. (2016). Concept of Advanced Thermal Barrier Functional Coatings in High Temperature Engineering Components. In A. Zuzuarregui & M. Morant-Miñana (Eds.), *Research Perspectives on Functional Micro- and Nanoscale Coatings* (pp. 396–419). Hershey, PA: IGI Global. doi:10.4018/978-1-5225-0066-7.ch015

Pau, J. L., Marín, A. G., Hernández, M. J., Cervera, M., & Piqueras, J. (2016). Analysis of Plasmonic Structures by Spectroscopic Ellipsometry. In A. Zuzuarregui & M. Morant-Miñana (Eds.), *Research Perspectives on Functional Micro- and Nanoscale Coatings* (pp. 208–239). Hershey, PA: IGI Global. doi:10.4018/978-1-5225-0066-7.ch008

Penchovsky, R. (2014). Nucleic Acids-Based Nanotechnology: Engineering Principals and Applications. In M. Bououdina & J. Davim (Eds.), *Handbook of Research on Nanoscience, Nanotechnology, and Advanced Materials* (pp. 414–430). Hershey, PA: IGI Global. doi:10.4018/978-1-4666-5824-0.ch016

Petrescu, L., Avram, S., Mernea, M., & Mihailescu, D. F. (2017). Up-Converting Nanoparticles: Promising Markers for Biomedical Applications. In M. Putz & M. Mirica (Eds.), *Sustainable Nanosystems Development, Properties, and Applications* (pp. 490–523). Hershey, PA: IGI Global. doi:10.4018/978-1-5225-0492-4.ch012

Philippe, A. (2017). Evaluation of Currently Available Techniques for Studying Colloids in Environmental Media: Introduction to Environmental Nanometrology. In S. Joo (Ed.), *Applying Nanotechnology for Environmental Sustainability* (pp. 1–26). Hershey, PA: IGI Global. doi:10.4018/978-1-5225-0585-3.ch001

Pirsa, S. (2017). Chemiresistive Gas Sensors Based on Conducting Polymers. In M. Ahmadi, R. Ismail, & S. Anwar (Eds.), *Handbook of Research on Nanoelectronic Sensor Modeling and Applications* (pp. 150–180). Hershey, PA: IGI Global. doi:10.4018/978-1-5225-0736-9.ch006

Popescu, L., Robu, A. C., & Zamfir, A. D. (2017). Sustainable Nanosystem Development for Mass Spectrometry: Applications in Proteomics and Glycomics. In M. Putz & M. Mirica (Eds.), *Sustainable Nanosystems Development, Properties, and Applications* (pp. 535–568). Hershey, PA: IGI Global. doi:10.4018/978-1-5225-0492-4.ch014

Pourasl, A. H., Ahmadi, M. T., Rahmani, M., Ismail, R., & Tan, M. L. (2017). Graphene and CNT Field Effect Transistors Based Biosensor Models. In M. Ahmadi, R. Ismail, & S. Anwar (Eds.), *Handbook of Research on Nanoelectronic Sensor Modeling and Applications* (pp. 294–333). Hershey, PA: IGI Global. doi:10.4018/978-1-5225-0736-9.ch012

Rafeeqi, T. A. (2015). Carbon Nanotubes: Basics, Biocompatibility, and Bio-Applications Including Their Use as a Scaffold in Cell Culture Systems. In M. Shah, M. Bhat, & J. Davim (Eds.), *Nanotechnology Applications for Improvements in Energy Efficiency and Environmental Management* (pp. 56–86). Hershey, PA: IGI Global. doi:10.4018/978-1-4666-6304-6.ch003

Rahmani, M., Karimi, H., Kiani, M. J., Pourasl, A. H., Rahmani, K., Ahmadi, M. T., & Ismail, R. (2017). Modeling Trilayer Graphene-Based DET Characteristics for a Nanoscale Sensor. In M. Ahmadi, R. Ismail, & S. Anwar (Eds.), *Handbook of Research on Nanoelectronic Sensor Modeling and Applications* (pp. 19–38). Hershey, PA: IGI Global. doi:10.4018/978-1-5225-0736-9.ch002

Rahmani, M., Rahmani, K., Kiani, M. J., Karimi, H., Akbari, E., Ahmadi, M. T., & Ismail, R. (2017). Development of Gas Sensor Model for Detection of NO_2 Molecules Adsorbed on Defect-Free and Defective Graphene. In M. Ahmadi, R. Ismail, & S. Anwar (Eds.), *Handbook of Research on Nanoelectronic Sensor Modeling and Applications* (pp. 208–223). Hershey, PA: IGI Global. doi:10.4018/978-1-5225-0736-9.ch008

Related Readings

Raniero, W., Della Mea, G., & Campostrini, M. (2016). Functionalization of Surfaces with Optical Coatings Produced by PVD Magnetron Sputtering. In A. Zuzuarregui & M. Morant-Miñana (Eds.), *Research Perspectives on Functional Micro- and Nanoscale Coatings* (pp. 170–207). Hershey, PA: IGI Global. doi:10.4018/978-1-5225-0066-7.ch007

Rodulfo-Baechler, S. M. (2016). Dual Role of Perovskite Hollow Fiber Membrane in the Methane Oxidation Reactions. In H. Al-Megren & T. Xiao (Eds.), *Petrochemical Catalyst Materials, Processes, and Emerging Technologies* (pp. 385–430). Hershey, PA: IGI Global. doi:10.4018/978-1-4666-9975-5.ch014

Rongione, N. A., Floerke, S. A., & Celik, E. (2017). Developments in Antibacterial Disinfection Techniques: Applications of Nanotechnology. In S. Joo (Ed.), *Applying Nanotechnology for Environmental Sustainability* (pp. 185–203). Hershey, PA: IGI Global. doi:10.4018/978-1-5225-0585-3.ch009

Roy, M. (2014). Nano Indentation Response of Various Thin Films Used for Tribological Applications. In M. Bououdina & J. Davim (Eds.), *Handbook of Research on Nanoscience, Nanotechnology, and Advanced Materials* (pp. 62–88). Hershey, PA: IGI Global. doi:10.4018/978-1-4666-5824-0.ch003

Sadeghi, H., & Sangtarash, S. (2017). Silicene Nanoribbons and Nanopores for Nanoelectronic Devices and Applications. In M. Ahmadi, R. Ismail, & S. Anwar (Eds.), *Handbook of Research on Nanoelectronic Sensor Modeling and Applications* (pp. 39–69). Hershey, PA: IGI Global. doi:10.4018/978-1-5225-0736-9.ch003

Saikia, P., Bharadwaj, S. K., & Miah, A. T. (2016). Peroxovanadates and Its Bio-Mimicking Relation with Vanadium Haloperoxidases. In M. Bououdina (Ed.), *Emerging Research on Bioinspired Materials Engineering* (pp. 199–221). Hershey, PA: IGI Global. doi:10.4018/978-1-4666-9811-6.ch007

Saladino, R., Botta, G., & Crucianelli, M. (2016). Advances in Nanotechnology Transition Metal Catalysts in Oxidative Desulfurization (ODS) Processes: Nanotechnology Applied to ODS Processing. In T. Saleh (Ed.), *Applying Nanotechnology to the Desulfurization Process in Petroleum Engineering* (pp. 180–215). Hershey, PA: IGI Global. doi:10.4018/978-1-4666-9545-0.ch006

Saleh, T. A., Danmaliki, G. I., & Shuaib, T. D. (2016). Nanocomposites and Hybrid Materials for Adsorptive Desulfurization. In T. Saleh (Ed.), *Applying Nanotechnology to the Desulfurization Process in Petroleum Engineering* (pp. 129–153). Hershey, PA: IGI Global. doi:10.4018/978-1-4666-9545-0.ch004

Saleh, T. A., Shuaib, T. D., Danmaliki, G. I., & Al-Daous, M. A. (2016). Carbon-Based Nanomaterials for Desulfurization: Classification, Preparation, and Evaluation. In T. Saleh (Ed.), *Applying Nanotechnology to the Desulfurization Process in Petroleum Engineering* (pp. 154–179). Hershey, PA: IGI Global. doi:10.4018/978-1-4666-9545-0.ch005

Salehi, M. A., Kadusarai, M. J., & Dogolsar, M. A. (2016). Capability of Bacterial Cellulose Membrane in Release of Doxycycline. *International Journal of Chemoinformatics and Chemical Engineering*, 5(1), 44–55. doi:10.4018/IJCCE.2016010104

Sareen, N., & Bhattacharya, S. (2016). Cleaner Energy Fuels: Hydrodesulfurization and Beyond. In T. Saleh (Ed.), *Applying Nanotechnology to the Desulfurization Process in Petroleum Engineering* (pp. 84–128). Hershey, PA: IGI Global. doi:10.4018/978-1-4666-9545-0.ch003

Shah, K. A., & Shah, M. A. (2014). Principles of Raman Scattering in Carbon Nanotubes. In M. Bououdina & J. Davim (Eds.), *Handbook of Research on Nanoscience, Nanotechnology, and Advanced Materials* (pp. 131–145). Hershey, PA: IGI Global. doi:10.4018/978-1-4666-5824-0.ch006

Shakir, I., Ali, Z., Rana, U. A., Nafady, A., Sarfraz, M., Al-Nashef, I., & Kang, D. et al. (2014). Nanostructured Materials for the Realization of Electrochemical Energy Storage and Conversion Devices: Status and Prospects. In M. Bououdina & J. Davim (Eds.), *Handbook of Research on Nanoscience, Nanotechnology, and Advanced Materials* (pp. 376–413). Hershey, PA: IGI Global. doi:10.4018/978-1-4666-5824-0.ch015

Sharma, P., Hussain, N., Das, M. R., Deshmukh, A. B., Shelke, M. V., Szunerits, S., & Boukherroub, R. (2014). Metal Oxide-Graphene Nanocomposites: Synthesis to Applications. In M. Bououdina & J. Davim (Eds.), *Handbook of Research on Nanoscience, Nanotechnology, and Advanced Materials* (pp. 196–225). Hershey, PA: IGI Global. doi:10.4018/978-1-4666-5824-0.ch010

Sharma, P., Hussain, N., Das, M. R., Deshmukh, A. B., Shelke, M. V., Szunerits, S., & Boukherroub, R. (2014). Metal Oxide-Graphene Nanocomposites: Synthesis to Applications. In M. Bououdina & J. Davim (Eds.), *Handbook of Research on Nanoscience, Nanotechnology, and Advanced Materials* (pp. 196–225). Hershey, PA: IGI Global. doi:10.4018/978-1-4666-5824-0.ch010

Related Readings

Shukla, R., Anapagaddi, R., Singh, A. K., Allen, J. K., Panchal, J. H., & Mistree, F. (2016). Integrated Computational Materials Engineering for Determining the Set Points of Unit Operations for Production of a Steel Product Mix. In S. Datta & J. Davim (Eds.), *Computational Approaches to Materials Design: Theoretical and Practical Aspects* (pp. 163–191). Hershey, PA: IGI Global. doi:10.4018/978-1-5225-0290-6.ch006

Souier, T. (2014). Conductive Probe Microscopy Investigation of Electrical and Charge Transport in Advanced Carbon Nanotubes and Nanofibers-Polymer Nanocomposites. In M. Bououdina & J. Davim (Eds.), *Handbook of Research on Nanoscience, Nanotechnology, and Advanced Materials* (pp. 343–375). Hershey, PA: IGI Global. doi:10.4018/978-1-4666-5824-0.ch014

Su, C., Puls, R. W., Krug, T. A., Watling, M. T., O'Hara, S. K., Quinn, J. W., & Ruiz, N. E. (2017). Long-Term Performance Evaluation of Groundwater Chlorinated Solvents Remediation Using Nanoscale Emulsified Zerovalent Iron at a Superfund Site. In S. Joo (Ed.), *Applying Nanotechnology for Environmental Sustainability* (pp. 92–111). Hershey, PA: IGI Global. doi:10.4018/978-1-5225-0585-3.ch005

Suar, S. K., Sinha, S., Mishra, A., & Tripathy, S. K. (2015). Fabrication of Metal@SnO2 Core-Shell Nanocomposites for Gas Sensing Applications. In S. Soni, A. Salhotra, & M. Suar (Eds.), *Handbook of Research on Diverse Applications of Nanotechnology in Biomedicine, Chemistry, and Engineering* (pp. 438–451). Hershey, PA: IGI Global. doi:10.4018/978-1-4666-6363-3.ch020

Sun, J., Wan, S., Lin, J., & Wang, Y. (2016). Advances in Catalytic Conversion of Syngas to Ethanol and Higher Alcohols. In H. Al-Megren & T. Xiao (Eds.), *Petrochemical Catalyst Materials, Processes, and Emerging Technologies* (pp. 177–215). Hershey, PA: IGI Global. doi:10.4018/978-1-4666-9975-5.ch008

Taborda, J. A., & López, E. O. (2016). Research Perspectives on Functional Micro and Nano Scale Coatings: New Advances in Nanocomposite Coatings for Severe Applications. In A. Zuzuarregui & M. Morant-Miñana (Eds.), *Research Perspectives on Functional Micro- and Nanoscale Coatings* (pp. 136–169). Hershey, PA: IGI Global. doi:10.4018/978-1-5225-0066-7.ch006

Torrens, F., & Castellano, G. (2017). Graphene and Fullenene Clusters: Molecular Polarizability and Ion–Di/Graphene Associations. In M. Putz & M. Mirica (Eds.), *Sustainable Nanosystems Development, Properties, and Applications* (pp. 569–599). Hershey, PA: IGI Global. doi:10.4018/978-1-5225-0492-4.ch015

Varahalarao, V., & Nayak, B. (2017). Microbial Nanotechnology: Mycofabrication of Nanoparticles and Their Novel Applications. In B. Nayak, A. Nanda, & M. Bhat (Eds.), *Integrating Biologically-Inspired Nanotechnology into Medical Practice* (pp. 102–131). Hershey, PA: IGI Global. doi:10.4018/978-1-5225-0610-2.ch005

Vargas-Bernal, R. (2015). Performance Analysis of Interconnects Based on Carbon Nanotubes for AMS/RF IC Design. In M. Fakhfakh, E. Tlelo-Cuautle, & M. Fino (Eds.), *Performance Optimization Techniques in Analog, Mixed-Signal, and Radio-Frequency Circuit Design* (pp. 336–363). Hershey, PA: IGI Global. doi:10.4018/978-1-4666-6627-6.ch014

Vargas-Bernal, R. (2016). Advances in Functional Nanocoatings Applied in the Aerospace Industry. In A. Zuzuarregui & M. Morant-Miñana (Eds.), *Research Perspectives on Functional Micro- and Nanoscale Coatings* (pp. 318–340). Hershey, PA: IGI Global. doi:10.4018/978-1-5225-0066-7.ch012

Vargas-Bernal, R. (2017). Modeling, Design, and Applications of the Gas Sensors Based on Graphene and Carbon Nanotubes. In M. Ahmadi, R. Ismail, & S. Anwar (Eds.), *Handbook of Research on Nanoelectronic Sensor Modeling and Applications* (pp. 181–207). Hershey, PA: IGI Global. doi:10.4018/978-1-5225-0736-9.ch007

Walker, G., Bououdina, M., Guo, Z. X., & Fruchart, D. (2014). Overview on Hydrogen Absorbing Materials: Structure, Microstructure, and Physical Properties. In M. Bououdina & J. Davim (Eds.), *Handbook of Research on Nanoscience, Nanotechnology, and Advanced Materials* (pp. 312–342). Hershey, PA: IGI Global. doi:10.4018/978-1-4666-5824-0.ch013

Wang, Z., Wu, P., Lan, L., & Ji, S. (2016). Preparation, Characterization and Desulfurization of the Supported Nickel Phosphide Catalysts. In H. Al-Megren & T. Xiao (Eds.), *Petrochemical Catalyst Materials, Processes, and Emerging Technologies* (pp. 431–458). Hershey, PA: IGI Global. doi:10.4018/978-1-4666-9975-5.ch015

Related Readings

Wani, I. A. (2015). Nanomaterials, Novel Preparation Routes, and Characterizations. In M. Shah, M. Bhat, & J. Davim (Eds.), *Nanotechnology Applications for Improvements in Energy Efficiency and Environmental Management* (pp. 1–40). Hershey, PA: IGI Global. doi:10.4018/978-1-4666-6304-6.ch001

Wani, I. A., & Ahmad, T. (2017). Understanding Toxicity of Nanomaterials in Biological Systems. In S. Joo (Ed.), *Applying Nanotechnology for Environmental Sustainability* (pp. 403–427). Hershey, PA: IGI Global. doi:10.4018/978-1-5225-0585-3.ch017

Wojtanowicz, A. K. (2014). Risk and Remediation of Irreducible Casing Pressure at Petroleum Wells. In D. Matanovic, N. Gaurina-Medjimurec, & K. Simon (Eds.), *Risk Analysis for Prevention of Hazardous Situations in Petroleum and Natural Gas Engineering* (pp. 155–180). Hershey, PA: IGI Global. doi:10.4018/978-1-4666-4777-0.ch008

Yasar, D., & Celik, N. (2017). Assessment of Advanced Biological Solid Waste Treatment Technologies for Sustainability. In S. Joo (Ed.), *Applying Nanotechnology for Environmental Sustainability* (pp. 204–230). Hershey, PA: IGI Global. doi:10.4018/978-1-5225-0585-3.ch010

Zarras, P., Goodman, P. A., & Stenger-Smith, J. D. (2016). Functional Polymeric Coatings: Synthesis, Properties, and Applications. In A. Zuzuarregui & M. Morant-Miñana (Eds.), *Research Perspectives on Functional Micro- and Nanoscale Coatings* (pp. 78–104). Hershey, PA: IGI Global. doi:10.4018/978-1-5225-0066-7.ch004

Zhang, T., Zhang, C., Xing, J., Xu, J., Li, C., Wang, P. C., & Liang, X. (2017). Multifunctional Dendrimers for Drug Nanocarriers. In R. Keservani, A. Sharma, & R. Kesharwani (Eds.), *Novel Approaches for Drug Delivery* (pp. 245–276). Hershey, PA: IGI Global. doi:10.4018/978-1-5225-0751-2.ch010

Zhizhin, G. V., & Diudea, M. V. (2017). Space of Nanoworld. In M. Putz & M. Mirica (Eds.), *Sustainable Nanosystems Development, Properties, and Applications* (pp. 214–236). Hershey, PA: IGI Global. doi:10.4018/978-1-5225-0492-4.ch007

Zuzuarregui, A., & Morant-Miñana, M. C. (2016). Functional Coatings: A Rapidly and Continuously Developing Field. In A. Zuzuarregui & M. Morant-Miñana (Eds.), *Research Perspectives on Functional Micro- and Nanoscale Coatings* (pp. 1–25). Hershey, PA: IGI Global. doi:10.4018/978-1-5225-0066-7.ch001

About the Authors

Hui Ge received his Ph.D. degree from Institute of Coal Chemistry, Chinese Academy of Sciences in 2009 and then worked there as an associate professor. He has been researching on synthesis, characterization, and catalytic application of Mo based materials and nanostructured composites. And He has developed series of CoMo and NiMo catalysts for hydrodesulfurization of gasoline and diesel. He has rich experience in synthesis of novel catalytic material and DFT calculation of the structure, properties and reaction mechanism of transit metal catalyst. His scientific interests include heterogeneous catalysis, 2D materials, theory calculation, new energy and environmental science.

Xingchen Liu is a research scientist at the Institute of Coal Chemistry, Chinese Academy of Sciences, China. He was born in 1984 in Shanxi Province, China. In 2002, he left for college education at Jilin University, where he got the degree of Bachelor of Science majored in chemistry in 2006, and the degree of Master of Science majored in physical chemistry in 2009. Afterwards, he studied in the field of theoretical chemistry under the supervision of Prof. Dennis Salahub at the University of Calgary as a graduate student. In 2015, after the Ph.D. graduation, he joined Dr. Xiaodong Wen's lab at the Institute of Coal Chemistry, Chinese Academy of Sciences and worked as a research scientist since then. Xingchen Liu's research focuses on the theoretical modelling of the heterogeneous catalytic materials at extreme conditions and complex environments. He also works on method developments for searching of reactions paths.

Shanmin Wang is a high-pressure (P) physicist and material scientist with extensive experience in diffraction studies of crystal structures, equations of state, phase transitions, and strongly correlated 3d systems, and hard/superhard ceramics. He has been working on synthesis, characterization, and modeling of catalyst materials, semiconductors and superconductors, hard/superhard materials, and nanostructured composites at high pressures. Prof. Wang has rich experience in use of high-P large-volume press and diamond-anvil cell. Very recently, using newly

About the Authors

formulated high-P reaction routes, he has recently synthesized a number of novel transition-metal nitrides, including W_2N_3, W_3N_4, and $3R-MoN_2$, and most of them are potential industrial catalysts possessing superior catalytic properties. Overall, his scientific interests include new energy researches, condensed matter physics, materials sciences, and crystallography. His technical specialties cover high P-T and high-P/low-T instrumentations for neutron and synchrotron x-ray diffraction.

Tao Yang received her Master degree from Lanzhou Institute of Chemical Physics, Chinese Academy of Sciences in 2005. In 2008, Dr. Tao Yang got her Ph.D. degree from Institute of Coal Chemistry, Chinese Academy of Sciences, and worked on Water-Gas-Shift reaction mechanism on iron oxides. After graduating, Dr. Tao Yang joined the group of Prof. Roald Hoffmann (the 1981 Nobel Prize in Chemistry) in Cornell University as a research scholar. In 2015, Dr. Tao Yang joined the group of Prof. Baojian Shen in State Key Laboratory of Heavy Oil Processing, China University of Petroleum as a research scientist, and is working on iron-based catalysts for hydrodesulfurization via theoretical approaches, including DFT and MD. Dr. Tao Yang's interests is to pursue and hunt a "route" to rational design Fe-based materials and catalysts with desired functionality.

Xiaodong Wen received his Ph.D. degree from Chinese Academy of Sciences in 2007. After his Ph.D., Prof. Wen joined the group of Prof. Roald Hoffmann (the 1981 Nobel Prize in Chemistry) and Dr. Neil Ashcroft at Cornell University as a postdoc. After three years, in 2010, he joined the group T-1 in LANL as a Seaborg Institute fellow, and worked on predicting actinide/magnetic materials using theoretical tools, and developing strongly correlated method. From 2014, Prof. Xiaodong Wen was selected as Hundred People Plan Program in Chinese Academy of Sciences, and awarded as National Thousand Young Talents Program of China. The researches of Prof. Wen are focused on rational design of catalysts (especially for carbon-based energy conversion) and materials (sustainable energy related materials) combining experimental and theoretical approach.

Index

A

active 9-11, 16-19, 24-26, 28, 31-41, 43-45, 47-50, 52, 59, 61-63, 65-67, 69, 71, 73, 75-80, 82-85, 90-93, 98-99, 101-102, 105, 110-111, 129-136
atom 8, 13-18, 39, 41, 106, 112, 114, 133

B

binding 14, 16, 19, 66, 114, 124

C

catalyst 9, 11, 18-19, 24, 26, 31-36, 38, 40-50, 58-59, 62-71, 73-87, 89-92, 94-96, 99, 101, 105, 107-108, 110-111, 114-115, 117-118, 120, 123-124, 126, 130-136
catalytic 1-3, 7-11, 13-14, 16, 18-21, 24-25, 28, 31-32, 35, 38, 40-41, 43-44, 46, 48-51, 53-54, 56, 60-61, 63, 67-68, 71-75, 77-80, 83-85, 87-88, 90-93, 98, 100-102, 105, 107, 111-112, 117-118, 121-122, 124-125, 130-133, 135-136
Chemical 2-9, 11-13, 16-17, 22-24, 26-29, 31-32, 35-36, 39-40, 42-43, 46, 48, 50, 54, 56, 58, 61-62, 64-68, 71, 73, 78, 81-86, 89, 91-94, 96-98, 100, 104, 106-107, 110, 113-118, 121-126, 128-131, 133-134, 136
CO2 9, 19-21, 43, 58, 68-69, 71-73, 75, 82, 84-85, 93, 112-114, 125, 129-134
crystals 3-5, 14, 16, 23, 61, 133
CVD 5-8, 66, 129, 133

D

dichlcogenides 1-3, 5-8, 11-14, 16, 18-21, 23, 27, 29, 53, 102, 121, 127, 129-134

E

electrochemical 3-4, 10-12, 18-19, 21, 24, 26, 29, 47, 92, 97, 130, 133-134
electron 9, 11, 14-16, 19-20, 24, 35, 61, 65, 71, 85, 102, 112, 114, 122, 128, 135
electron-hole 10, 19
electronic 3, 8, 11, 16, 18, 24-25, 32, 37, 88, 101-102, 105, 118, 121, 123-124, 127, 130, 133, 135-136
Engineering 8, 14, 25, 29, 45-46, 50, 86, 125-126, 129, 133, 135-136
evolution 9, 16-19, 21-22, 24-26, 28-30, 47, 65-66, 69, 73, 84-85, 87, 90-93, 95-98, 133-134
exfoliation 3-5, 7, 11, 22-24, 27-28, 30, 118-119, 130, 133

F

fabricated 3, 8, 70

G

gas 3-5, 19, 43, 48, 55, 60-61, 68, 72-73, 77-78, 83, 90, 93-94, 130, 134

H

heterostructure 2, 7-10, 14-15, 23, 25, 102, 120, 130, 133

Index

hydrogen 3, 5, 9, 16-19, 21-22, 24-26, 28-30, 35, 52, 54, 59, 62, 65-71, 76, 79, 84-85, 90-98, 107, 110, 112, 120, 126, 129, 133-134

I

intercalation 2-4, 11-12, 130, 133
interlayer 3-4, 10, 23, 102

K

kinetics 18-19, 33, 38, 75, 121, 126

L

lattice 8, 12, 37, 60, 77, 111, 130
layer 1, 3-8, 10, 19-21, 26, 31-32, 59, 63, 65-66, 70, 85, 90, 106, 112, 129, 133, 136
liquid 4-5, 19, 21-23, 27, 46, 72, 96, 119, 134
lithium 3-4, 11-12, 26

M

material 1-5, 7-8, 10-12, 14, 16-17, 21-25, 27-29, 32, 37, 44, 53-55, 58, 65, 68-69, 71, 75, 77-78, 80, 84, 86-88, 90-93, 96-98, 100-102, 111, 117-121, 125, 127, 129-135
metal 1, 3, 7, 12, 14, 18, 21-30, 37, 45-46, 49, 55, 58-60, 62, 66, 68, 70-73, 75-76, 79, 83-85, 87, 89-90, 92-94, 96-97, 102, 111, 117, 119-125, 127, 131, 134
metallic 3, 14, 16, 18, 21, 26, 37, 48, 69, 101-102, 111-112, 118
methods 2, 4-5, 7, 9, 14, 16, 45, 56, 60, 72, 89, 115, 130, 135
microscopy 14-15, 48, 52, 65, 87, 112, 128, 135
Molecular 11, 18, 23, 25, 28-29, 48, 50, 58, 77, 88-91, 93, 96, 98, 105-107, 110, 112, 116, 118-120, 123-124, 126, 128-129

molybdenum 19, 21, 23-25, 27-28, 30, 40, 43, 49, 54-60, 62-63, 66-69, 71, 74-78, 80-84, 86-99, 101, 111-112, 116-127, 131, 133, 135
monolayer 1, 9, 14-16, 18, 25, 27-28, 63, 101-102, 104-105, 117-118, 121, 123, 127
Monolayers 5, 14, 27, 29, 120, 125
MoS_2 1, 3-5, 7-13, 16-29, 32, 34-35, 37-39, 41, 43-48, 50-52, 61, 70, 96, 101-102, 104-108, 110, 117-127, 133, 135
MoS_x 9, 18-19, 39, 70, 126
Mo(W) 1-8, 10-13, 16, 18, 20-21, 43, 53-54, 56, 58, 60, 62, 66-67, 75, 82, 129-134

N

nanoparticles 9-10, 16-17, 19-20, 24, 26, 29-31, 38, 40, 54, 56-59, 63, 66, 74, 76, 80, 83, 85-87, 92, 94-95, 98, 100-101, 105, 107-108, 112, 115-116, 122-123, 125
nanosheets 1-8, 10-13, 18-19, 22-24, 26-30, 32, 36, 69, 84, 119, 133
nanowires 10, 18, 22, 56, 64-65, 69, 72-73, 86, 90, 93, 97

O

octahedral 11, 13, 18, 63, 111
overpotentials 18-20, 70
oxidation 10-11, 22, 61, 63, 66, 68, 73-80, 82-83, 86, 88-90, 94-97, 99, 108, 120-122, 125, 131-132, 134
oxide 9-10, 33, 53-54, 59-60, 62-66, 71, 73-90, 92-94, 96, 98, 120, 131-132, 134

P

phase 3-6, 11-14, 18, 26-27, 32, 34-37, 39-41, 43, 47-49, 51-52, 54, 64, 68, 71-72, 74-75, 77-78, 82-83, 86, 88, 90, 92, 96, 117, 123, 130, 132

photocatalytic 10, 19, 22, 29-30, 72-73, 87-88, 97
polymer 3, 58, 75, 132
properties 1-4, 8, 10, 14, 20, 24-25, 27, 30, 32-34, 41, 46, 49, 53, 63, 70, 76, 80, 88, 96, 100-102, 105, 111-112, 118, 120-124, 130-131, 133, 135-136

R

reaction 2-3, 7-10, 12, 16-19, 25-26, 28-31, 33-38, 40, 42-43, 45-46, 48, 54, 58-62, 65-66, 68-69, 72-74, 76-81, 84-85, 87, 90-99, 108, 110, 116, 124, 131-136

S

semiconductors 8, 23, 70, 127
solvent 3-4, 7, 23, 33, 76, 116, 130
solvothermal 7, 9, 76
spectroscopy 14, 16, 51, 63, 80, 101, 105
structure 1, 8, 10-13, 16, 24-25, 27, 32-33, 36-37, 40, 48, 50, 52, 54, 57, 63-64, 73-75, 84, 91, 96-97, 99, 101-107, 111-112, 117-119, 121-122, 125-126, 130, 133, 135-136
superlattices 7

surface 4-5, 7, 10, 17, 19, 25, 32-35, 37, 39, 54, 56-60, 63, 65-66, 68, 70, 74, 76-80, 83, 88, 97, 102, 105, 107-108, 111-112, 114-116, 118, 121, 123-127, 134-135
surfactant 3, 5, 7, 33, 63, 130
symmetry 3, 13-14, 32, 94, 117
synthesis 1, 3, 9-10, 21, 23-27, 33, 43-44, 46-51, 53, 56, 59-63, 66, 71-73, 76, 79, 84-90, 92-94, 96-98, 111, 121, 131

T

thermodynamically 8, 13, 37, 60, 105, 107-108
transmission 14-15, 112, 128, 135
tungsten 9, 19, 40, 54, 60, 65, 70-71, 74, 78, 80, 83, 85, 87-88, 90, 93-94, 96-97, 122

U

ultrasonic 3

V

vapor 3, 5-7, 28, 66, 73
vertical 6, 8, 14, 25, 120, 133

Purchase Print + Free E-Book or E-Book Only

Purchase a print book through the IGI Global Online Bookstore and receive the e-book for free or purchase the e-book only! Shipping fees apply.

www.igi-global.com

Recommended Reference Books

ISBN: 978-1-4666-7387-8
© 2015; 994 pp.
List Price: $412

ISBN: 978-1-4666-5824-0
© 2014; 617 pp.
List Price: $292

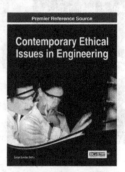

ISBN: 978-1-4666-8130-9
© 2015; 343 pp.
List Price: $172

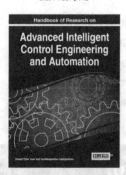

ISBN: 978-1-4666-7248-2
© 2015; 794 pp.
List Price: $268

ISBN: 978-1-4666-7320-5
© 2015; 343 pp.
List Price: $172

ISBN: 978-1-4666-5800-4
© 2014; 556 pp.
List Price: $188

*IGI Global now offers the exclusive opportunity to receive a free e-book with the purchase of the publication in print, or purchase any e-book publication only. You choose the format that best suits your needs. This offer is only valid on purchases made directly through IGI Global's Online Bookstore and not intended for use by book distributors or wholesalers. Shipping fees will be applied for hardcover purchases during checkout if this option is selected.

Should a new edition of any given publication become available, access will not be extended on the new edition and will only be available for the purchased publication. If a new edition becomes available, you will not lose access, but you would no longer receive new content for that publication (i.e. updates). The free e-book is only available to single institutions that purchase printed publications through IGI Global. Sharing the free e-book is prohibited and will result in the termination of e-access.

Publishing Information Science and Technology Research Since 1988

www.igi-global.com   Sign up at www.igi-global.com/newsletters   facebook.com/igiglobal   twitter.com/igiglobal

Stay Current on the Latest Emerging Research Developments

Become an IGI Global Reviewer for Authored Book Projects

The overall success of an authored book project is dependent on quality and timely reviews.

In this competitive age of scholarly publishing, constructive and timely feedback significantly decreases the turnaround time of manuscripts from submission to acceptance, allowing the publication and discovery of progressive research at a much more expeditious rate. Several IGI Global authored book projects are currently seeking highly qualified experts in the field to fill vacancies on their respective editorial review boards:

Applications may be sent to:
development@igi-global.com

Applicants must have a doctorate (or an equivalent degree) as well as publishing and reviewing experience. Reviewers are asked to write reviews in a timely, collegial, and constructive manner. All reviewers will begin their role on an ad-hoc basis for a period of one year, and upon successful completion of this term can be considered for full editorial review board status, with the potential for a subsequent promotion to Associate Editor.

If you have a colleague that may be interested in this opportunity, we encourage you to share this information with them.

Become an IRMA Member

Members of the **Information Resources Management Association (IRMA)** understand the importance of community within their field of study. The Information Resources Management Association is an ideal venue through which professionals, students, and academicians can convene and share the latest industry innovations and scholarly research that is changing the field of information science and technology. Become a member today and enjoy the benefits of membership as well as the opportunity to collaborate and network with fellow experts in the field.

IRMA Membership Benefits:

- **One FREE Journal Subscription**
- **30% Off Additional Journal Subscriptions**
- **20% Off Book Purchases**
- Updates on the latest events and research on Information Resources Management through the IRMA-L listserv.
- Updates on new open access and downloadable content added to Research IRM.
- A copy of the Information Technology Management Newsletter twice a year.
- A certificate of membership.

IRMA Membership $195

Scan code or visit **irma-international.org** and begin by selecting your free journal subscription.

Membership is good for one full year.

www.irma-international.org

Printed in the United States
By Bookmasters